THE COMPLETE GUIDE TO SMALLHOLDING

Debbie Kingsley

THE CROWOOD PRESS

First published in 2023 by
The Crowood Press Ltd
Ramsbury, Marlborough
Wiltshire SN8 2HR

enquiries@crowood.com

www.crowood.com

British Library Cataloguing-in-Publication Data
A catalogue record for this book is available from the British Library.

ISBN 978 0 7198 4215 3

Typeset by Envisage IT
Cover design by Blue Sunflower Creative
Printed and bound in India by Replika Press Pvt Ltd

Contents

1 Introduction

COMING CLEAN

If you're looking for a bible for complete self-sufficiency, this isn't it. Just the idea of knitting my own toilet roll or roasting dandelion roots to create a coffee substitute to succour my friends makes me chuckle. I'm not interested in eating cabbage at every meal because it might grow well in our soil, or eating slices from a loaf that sounds, tastes and looks like a brick.

No one could ever give me the epithet of being worthy. I want to eat asparagus cut fresh, moments before the spears go in the pan, and suck raspberries off each finger in greedy glee. I also want a store of stunning beef in the freezer to make a meal of sirloin steak accompanied by field mushrooms gathered that morning and peas fresh from the pod, or a thick slice of gammon glazed with mustard and honey alongside a duck egg pillowed on mayonnaise and scattered with chives with newly dug salad potatoes. I want (I want a lot, don't I?) vats of cider glugging away in the store room for drinking and for cooking with rare-breed pork with chunks of apple in a casserole, and for turning into cider vinegar that will mutate into blackberry vinegar to accompany salad leaves grown in the polytunnel.

I want, when all's said and done, to grow, rear and make delicious things and have an interesting, seasonal, nature-observing and enhancing way of life.

There are so many different ways of smallholding and being a smallholder, and it should not be a way of life that makes a rod for your own back. Just because someone in the next village or on one of the ubiquitous rural life television programmes finds joy in keeping rabbits for the pot, or grows an acre of wheat that they thresh by hand and mill for flour, doesn't mean you have to. You might yearn for charcuterie without nitrates and have a fancy for pig-keeping, love eating chicken but have a fear of birds (in that case buy oven-ready quality ones from another source and don't keep poultry), or have a passion for jams and chutneys and be keen to grow fruit and vegetables.

This way of life should not be about wearing a hair shirt as if it only feels real if there's some suffering in the mix. Stuff that: life is hard enough. If producing the majority of your own food is an all-absorbing ambition, that's something we share; you have so many possibilities ahead of you, so choose the things that bring you joy.

What I hope this book will give you is a comprehensive insight and guidance into the various elements of a possible smallholding life. Pick and choose the bits that feel right and fit with your life and aspirations, and ignore the parts that don't (although you can't ignore everything – if you have livestock or make food for sale there are rules and regulations).

The reason that smallholding is of abiding interest for so many people is precisely because it offers a different way of being, away from the career ladder, commuting and office politicking. Being in charge of every decision is hugely freeing, but the reality is that this way of life has significant ties: this is a 365-days-a-year job (although you can have time out and holidays – we'll get to that bit later). It's an opportunity to learn a lot of new things and new ways of doing things, not re-creating the irritations, furies and sadnesses of office life. And because making this way of life pay is something that so many potential and new smallholders want to know about, there's plenty of grounding, pragmatic information about that too.

A smallholding idyll.

THE GOOD SMALLHOLDER

There are certain traits that undoubtedly help make a smallholder. Having a robust constitution, both physical and mental, is undoubtedly helpful, although it's also clear that smallholding activities can help heal a troubled mind. Machinery is a life-saver when the muscles age and tire or if you have limited strength, but there is no denying that there is a great deal more physicality required in this life than in many others. Some days I can haul a 25kg sack of pig feed over my shoulder and trudge up the track with it, on other days it requires the help of a wheelbarrow.

Having self-discipline is crucial when you have livestock; those chitting potatoes might be able to wait another week before being planted out, but your livestock have to be fed, watered and checked every single day without fail, usually before you contemplate your own breakfast. If poo in whatever form freaks you out, you'll need to overcome this. Dealing with mortality becomes very real. If, like me, rats make you squeal (this hasn't improved much in thirty years), you learn to squeak and carry on regardless.

Having an aptitude for, or at least no fear of learning how to make and mend things, whether that is a fence or a goat shed, is helpful. Using hand and power tools is something you're just going to be doing. You might start off hammering the fencing stake ten times before you manage to hit the staple, but you'll get there if you hold the hammer properly and don't close your eyes while you do it.

Having project management skills is surprisingly useful; smallholding is nothing if not a raft of multiple projects with seemingly simultaneous demands on your time, and no boss other than the seasons providing a critical path analysis.

Smallholding is great for the immensely practical person, but it's also a thrill for those who love a mental challenge: there is so much that you can learn, from land management and enhancement, animal husbandry, midwifery, disease diagnostic skills, fencing, breeding, meat production, using tools effectively – and on and on it goes. Smallholding can keep you intellectually challenged and engaged for life. Unless you come from a farming or veterinary background, the skills and knowledge that you need are not something that most of us learned during our formal education, and there will be a whole swathe of new terminology and confusing equipment to get your head round.

Never has observation and action-based learning been more important as when a life depends on you, whether it be a duckling hatching out in an incubator, or at lambing time. You can (and should) go on some of the excellent courses now available to give you a kickstart and a proper grounding on topics where your enthusiasm outweighs your knowledge by the power of ten. But don't hop continually from one course to the next without putting into practice some of your new understanding: there comes a point when you just have to get stuck in.

Pilgrim geese in the farmyard.

Learn to use your eyes, ears, nose and hands around your new livestock. My nose alerts me immediately to any case of flystrike if we're handling the sheep – and more importantly, reassures that we *don't* have any problems of that nature. My ears will tell me if a sheep is in distress several fields away – a head stuck in a fence perhaps, or a lamb separated from its mother. Eyes have to be on full alert – you don't simply count the hens, lambs and pigs in the morning when you do your rounds, you are looking, always, for things that you don't want to find, so you can intervene early and avoid more serious complaints caused by failing to notice problems. And get hands on: a thick fleece or heavy feathering can hide poor body condition or parasites. Know how heavy your hens should feel when you pick them up, and how prominent the spine is on a well fed or an underweight sheep.

In the first few years, absolutely everything is a new challenge, from how to put up a livestock-proof fence, to what to feed your ducks, what works best as bedding in your poultry huts, and how to grow vegetables when you have heavy clay soil. But with the basics sorted, and with growing experience and a keen mind, your learning moves on to more demanding issues – to more effective land management, dealing with health issues without constant recourse to the vet, and breeding.

Then there's managing your plot for encouraging wildlife and native flora, rainwater harvesting, drainage, and possibly tractor and machinery use and maintenance. If you are scientifically minded you can explore carbon sequestration and soil health, do your own faecal egg counts and parasite analyses under a microscope, assess the impact of different grazing approaches on your land, and investigate the effect that minerals have on the soil and livestock. For those of a creative bent, working with fleece from the many and varied native breeds of sheep, or carving homegrown timber may satisfy the desire to make beautiful objects, and you will have endless subjects to inspire you if your interest lies in photography, painting or drawing.

You can develop the skills and understanding of an ecologist, environmentalist, nutritionist, agronomist and botanist. You can join schemes to monitor and maintain the health of your livestock, and the improvement of their conformation. If you are interested in showing your stock, that's a whole other learning curve, and would satisfy the most competitive smallholder urge. You could learn about and follow organic principles, explore the pros and cons of raw milk consumption, or write children's stories about a pet sheep named Curly.

Some smallholders become passionate about chickens – they keep many breeds, really understand the genetics behind feather colour, and can talk all things hen for hours on end – whereas I just want big, easy birds for delicious meat. Some smallholders have a passion for sheep and either keep one breed until they get as close to their idea of perfection as possible, or keep many different breeds to satisfy the urge for experimentation with fleeces or to provide a wonderful, varied view out of their bedroom window. As for me, I'm happy chatting about cows all day long. The point is that we all have our passions, and if you choose smallholding as a way of life you will have plenty of potential interests to enthuse about in the best possible way.

However, do avoid the known danger of acquiring something of everything and failing to acquire the knowledge you need to keep it all in good heart. It's also worth saying that for the many who successfully take on a couple of weaners to rear for pork and bacon, additional knowledge is required to breed a sow to produce your own piglets: don't underestimate the difference between rearing and breeding livestock. In particular, knowing what you're going to do with the resulting livestock or meat needs planning.

The experience you gain over time may all be directed towards the improvement of your smallholding practices, or it might be tangential, using the plant and animal life as inspiration for other pastimes. What I *can* tell you is that there's much to learn, much to enjoy, and much to hold your interest over a long lifetime.

More than anything, a 'can-do' attitude will get you through many demanding tasks, and the resulting sense of achievement is unassailable. In the winter our cows are housed for five months and we muck them out every single day, which becomes tiring by the time spring turnout approaches – and yet I still close the cowshed gate and turn to watch them munch on their haylage and smile at the clean concrete and freshly fluffed straw bedding with pleasure.

DEFINITIONS OF A SMALLHOLDING

Definitions of what a smallholding is range from the simple 'a small farm' to the equally vague 'an area of land that is used for farming but is much smaller than a typical farm', to the strangely specific, if out of date 'land acquired by a council that exceeds one acre and either does not exceed fifty acres or is of an annual value not exceeding fifty pounds' (at 1933 prices). A smallholding may simply be a patch of land where something agriculturally productive is happening.

Smallholders have to put up with many labels, some of which aren't always applied entirely kindly. Hobby farmer is one such term, intimating someone playing at having a few animals and, it's inferred, really making an amateurish hash of things. And then there's the patronising 'good-lifer', as if the smallholder is shrugging off the benefits of modern life and living on dandelion wine, kale and scrawny chickens. I have far more respect for smallholders than those terms imply, and define smallholding in a number of ways – but the main principle is using whatever means and resources you have available to you to produce a quantity of your own food. That may be more than growing a few herbs in a window box or a dozen radishes and a row of lettuces in the back garden, or perhaps not.

And who is to say that this food production should involve animals of any sort? You may be growing an orchard that provides you with fruit and nuts, and cider and vinegar, and wine and fabulous puddings, or a vegetable garden that focuses on luxury vegetables such as salad potatoes, asparagus, artichokes, chillies, aubergines and crisp, fresh sweetcorn, and a fruit cage with blueberries, whitecurrants and autumn raspberries. Or you may have a pragmatic vegetable patch where your harvest of spuds, carrots, onions, peas and brassicas feeds the family all year round.

Home-grown breakfast.

An interest in food and food production is a critical element of smallholding – growing or rearing food for its superior flavour, nurtured for its eating quality and not jammed full of chemicals. The driving force behind many smallholders is to put great quality food of known provenance on the table, and there is no better provenance than home grown. There is also the element of producing many different things, for who wants to eat the same meal every day of the year?

Does smallholding require access to land? I'd say yes, but this could be a vigorously used allotment or a community-owned plot: who is to say that land sharing rather than personal ownership makes you a real smallholder? Not me. And I've known plenty of smallholders who have been more productive in a large suburban garden than those with a 5-acre pony paddock.

Does Size Matter?

Size does matter. Not that you aren't a proper smallholder unless you have x amount of land, but because you have to take into account the scale of what you have to work with when making choices about what you will do with your space. You can't put six cows, twelve goats and ten sheep on a single acre of land – well, you could, but it would be the most unholy mess of disease, and worse.

FROM SUBSISTENCE FARMING TO LIFESTYLE CHOICE

Subsistence farming is the growing and rearing of enough food to feed your own family to avoid starvation, with, if you are lucky, a small cash-crop surplus to trade for other goods. Although it may not be common in the developed world these days, sub-Saharan Africa and parts of South America have many subsistence farmers. There are about 500 million small farms in developing countries, supporting almost two billion people – one-third of humanity. Agriculture on the micro scale has existed almost as long as humans (certainly from 8500BCE), and there have been dramatic developments over time in how the world farms.

But looking at smallholding in particular there has been a noticeable evolution in recent times regarding what it means to be a smallholder. We have moved from the necessity of subsistence farming in order to stay alive, to widening our dietary options on a tight budget, to making a lifestyle choice to be more self-sufficient, and yet further, to rearing livestock as desirable pets for those with no budgetary constraints.

It may seem strange that a peasant life has become a desirable model for those with shallow or deep pockets, but nature, seasonality, fresh food and clean air are fundamental to our wellbeing, and smallholding offers that day in, day out.

Biophilia

Humans are innately attracted to the natural world. Even the most urban soul enjoys a stroll through the park, and for many, being in the open air, away from high rise buildings and the constant roar of traffic, is an essential part of maintaining a healthy mind and body. Smallholding takes the weekly hike through the nearest slice of green belt and the Saturday stint at the allotment into another realm of commitment. In exchange, your biophilia will be well catered for.

Giving Life Focus

We live in challenging times in a world being depleted, ravaged and disrespected. Contemplating this can be overwhelming at times, and although there are many things that an individual can do, there is so much that is also out of our specific control. Smallholding can give life a positive focus, concentrating on the things we can do and grow and create that improves things for those around us, including livestock and wildlife, and for the elements that surround us: water, soil and air.

This book isn't a diatribe against the huge corporations that spend billions trying to convince us that laboratory-produced 'food' is better for us and our planet (I prefer to apply myself to what I can change, and others are better qualified than I to do the articulate railing against greed that's needed) – but I know for certain that our home-grown 100 per cent pasture-fed beef and lamb is not the problem that needs tackling, and is in fact part of the solution. You may find that a smallholding life gives solace, and a sense that you can control your small slice of the world.

Way of Life or Lifestyle?

Only you (and your bank account) can decide if smallholding will be a full-time occupation, a focus for a busy retirement, a wonderful way to spend your weekends and evenings, or shared with other part-time working or as an element of a portfolio career. The sale of a moderate city house can still buy a place in the country with accompanying land, and often the land is an important part of the equation. In some cases, the realisation comes later that something needs to be done with the newly acquired acreage, and so a spate of accidental smallholding starts, and sometimes flourishes, sometimes not.

Hobby or Business?

What about the commercial aspect? Do you need to sell surplus sausages or eggs to friends and neighbours to be a smallholder? No. Do you need to make an income from other activities to be a smallholder? That will be dictated by your financial situation and life plan. If your smallholding is a business, does that stop you being described as a smallholder? No, of course not – making a financial success of this way of life is something many aim for and few attain.

Self-Sufficiency, or a Contribution to the Table?

A contribution to the table is what most smallholders strive for, whether that be focused on creating the full regalia of the Christmas dinner and perhaps a hog roast for friends in the summer, to producing the vast majority of everything that finds its way on

Produce stall with honesty box.

to our plate. It's such a personal thing, wrapped up as it is in acreage, time, interests and personal drive.

We produce all our own beef, lamb, hogget, mutton, pork, bacon, sausages, gammon, duck, goose, goat, eggs, much of our chicken, apple juice, cider, blackberry and cider vinegars, chillies, and a vast array of veg in season plus orchard fruits – and away from the table we produce all our firewood, harvest rainwater, and get electricity from the sun. We don't produce our own milk or honey or cheese (cheese is my retirement plan). We make heaps of jam, chutneys and relishes, because we love and use them. I don't knit or sew my own clothes, although I used to do lots of both. Have a go at many things, and stay with those that give you pleasure or are necessities.

Pets or Livestock?

What about people keeping livestock as pets? Does having a few graceful grazing alpacas make you a smallholder, or only if you keep them to see unwanted foxes off your poultry and lambs? Is a mini private petting zoo a smallholding? Things get a little blurry here, there being no food production element involved. If your animal keeping is more akin to having a couple of ponies, perhaps you don't see yourself as a smallholder at all. However, an increasingly popular activity is the rearing for sale and keeping of livestock as pets.

This is a far cry from the original smallholding ethos, where everything produced was for consumption, by the producer or their customer. This is an indicator of a moneyed society, where some can indulge in statement livestock that enhance the look and feel of a country home. In the past this role was carried out by peacocks and herds of deer, but pet pigs, camelids, sheep, cattle, wallabies and emus now show themselves off in pastures that are put to no use other than fun, pleasure and rescue.

Part of the drive for livestock as pets is an innate desire to get back to nature and be closer to the land, even if taking the animals off to the abattoir is a step too far. Of course, the care of such creatures is no less demanding than those raised for meat.

Forever Popular

Taking back a significant degree of control over your own life is a huge comfort in uncertain times: when things are tough you might not be able to buy a pair of new shoes, but there's always meat in the freezer, and veg and fruit in the garden. Creating a better life for family and children, with a wider appreciation of the natural world, plays a strong part in stimulating the smallholder.

OUR STORY

We all come to smallholding in different ways, and ours was accidental. In our early twenties and looking for a first home together without housemates, we came across a small converted dairy attached to a farm and riding school. We took one look at the tiny dark rooms with a bedroom in the eaves and ancient woodburning range, and said yes. The farmer renting out the place looked doubtful, as the last folk from London had lasted a week – but seven years later we had to wrench ourselves away. We got involved in pretty much everything on the farm: we got a horse, milked goats and cows, sheared sheep, built stables, created a veg garden, foraged for blackberries and field mushrooms, helped with fencing, made hay and the best friends ever.

Eventually we bought our own smallholding – a farmworker's cottage with 3 acres that included a small barn, workshop, mature orchard, a stream for ducks and geese and a field. Wielding the slash hook, we unearthed and then restored duck huts, a chicken coop and run. Then we put up a greenhouse, created a big veg garden, and bought our first pigs and sheep.

A few weeks after arriving we watched a neighbour drive up and down their 3 acres on a sit-down mower, stopped to chat and asked if he enjoyed doing that. He didn't, and we took over his acreage as sheep pasture in exchange for a lamb carcase each year – possibly the cheapest rental ever. We kept guinea fowl, chickens, geese and ducks, and were adopted by a stray peahen who left in a grump when we got her a peacock mate. After a ghastly dog attack on our sheep we bought a llama, which kept foxes and dogs at bay.

We ran the smallholding alongside demanding jobs. Andrew spent quite some time working abroad, and working in the arts I frequently worked evenings as well as the usual working week. We had to be very well organised and have fences that kept everything safely in and predators out, and took our holidays at lambing. In many ways it was the idyllic set-up, but we had energy to spare and wanted a bigger challenge, and ten years on started hunting for a 20- or 30-acre property. A quiet environment was absolutely top requirement, and we eventually found a small farm in West Devon with thirteen tumble-down buildings, dubious fencing, run-through hedges and over a hundred acres – but the house was in a decent if basic state of repair, and the location wonderful.

The cob barn before.

The cob barn after.

Our farmer friends took one look and went quiet, but by then we'd exchanged contracts and somehow weren't fazed. We thought it would be a twenty-five-year project, but the majority of the work was achieved in less than ten. To start with we used 14 acres and rented out the rest to a wonderful farming family who have become great friends and mentors, and took back more land over time as our smallholding ambitions grew.

With twenty years of experience behind us, in 2009 we started to run smallholding courses of the kind we wished we'd had access to in the beginning. They have been phenomenally successful, and the people we meet come from all over the world; it has given us a great deal of joy sharing our learning, and hearing what next steps people take in their smallholding journey. With well over thirty years of smallholding experience, this book is a bringing together of the information we share on our courses, and a great deal more.

Reality Check

This book is fairly chunky, but it would have extended to fifty volumes if it included every smallholder possibility. Continue to enhance your knowledge, seeking out specific areas of interest, from home butchery to beekeeping, salami making to living off-grid. Smallholding can be a lifelong journey of acquiring information and skills.

2 Preparing for the Smallholding Life

Not everyone has to move house – or can afford to – in order to become a smallholder. You may have a garden big enough to keep a few hens, a trio of ducks, some rabbits, a few fruit bushes, window boxes of herbs and a veg patch. Or perhaps there's a neighbour with a large garden that's rather neglected that you could use in exchange for some fresh produce, or an allotment with your name on it. None of these options disrupts your life, and any one of them is practical, inexpensive and rewarding. But if your ambitions are broader, or your current situation provides minimal opportunities for testing out the smallholder existence, there are other ways to try before you buy into the whole smallholder way of life.

Why bother? Why not just jump into it with glee? In reality, moving home is an expensive process, and you may have to move to a completely new location to pursue your dream, with the loss of friends, neighbours, facilities, schools and employment opportunities that this entails. I estimate that more than 95 per cent of people who come on our smallholding courses are keen to get going, and after a couple of days they want to bring their plans forwards if at all possible and make a start. But a small yet noteworthy number of participants come up to me at the end of a course and thank us profusely, and say that although they've had a great time, they now know absolutely that it's not the life for them (they hadn't realised it was such hard work, that they would be responsible for doing this or that task, or that it was really not a good idea just to buy a few sheep, leave them in a field and let them get on with it).

TRY BEFORE YOU BUY

Trying before you buy is all about taking a reality check and looking at things sensibly. Even a day or two of experience or training can be enough to convince you that either this is truly your dream existence, or that there is absolutely no chance whatsoever of having, say, pigs, or even living in a rural location. Save yourself the grief and expense of an inappropriate purchase by first finding out what it involves – and you don't need to throw in your job and volunteer with a smallholder in an inaccessible part of the country to get a good feel for it.

Smallholding course participants.

The first step, which won't even get you out of your armchair, is to read a book or two. There are plenty around (*see* the appendices for recommendations), from in-depth studies of a specific topic (goats, lambing, bees or permaculture, for example) to lifestyle tales that depict a year in the life of a smallholder, and broader, comprehensive works such as this one that give you enough information to really understand what's involved. Then there's the world of free on-line videos, sharing skills such as artificially inseminating a pig to clipping a hen's wing, with accounts that range from hilarious, to fascinatingly informative, to terrifyingly unsafe or inappropriate.

Increasing numbers of holiday cottages offer some smallholder experience: collecting the eggs, bottle feeding lambs, mucking out the pigs and so on, all of which can be an enjoyable way of getting the whole family involved in a light-hearted manner with minimum responsibility attached. You might have a local smallholder who would love a friendly hand from time to time, or you could volunteer at a city farm. With a stretch of free time, you could try one of the farm volunteer programmes, such as HelpX, Wwoof, WorkAway or Volunteers Base, which have farm and smallholding hosts all over the world.

For a day or two of focused training there are a number of smallholding courses around the country, including our own. Make sure that what's on offer suits what you need, and that the trainers have plenty of experience in both delivering training and running a smallholding – it was attending a truly appalling training day that encouraged us to create courses we thought would be of real value to participants.

PUTTING NEW SKILLS INTO PRACTICE

You could read every book on the subject and go on all the courses on offer, but with some sense of the realities now in place, ultimately there will be nothing as nerve-rackingly rewarding as putting those new skills into practice and learning on the job. For example, it's incredibly useful (actually, it's probably essential) to be shown how to handle a sheep, and to tackle some yourself under supervision – but the practice you'll get by dealing with your own flock will build on that learning and develop your skill to a level where you become confident in your own abilities – and there are no shortcuts for that.

I do want to raise a warning flag about the advice given so freely on social media. There are some extremely knowledgeable people on these platforms happy to share information and to respond to questions from the worried new smallholder. The trouble as a learner is knowing the difference between the good stuff and some of the awful suggestions made, which if followed could cause damage, or worse, to the land or animals under discussion. Even the most experienced can give poor guidance because they are swept away by the immediacy of social media interactions – so don't, for example, ask Facebook for the dosage of certain medications or their withdrawal period, or what home remedies are cheaper than the medication that has been prescribed for your poorly pig by your (expert) vet.

MAKING FRIENDS

There are several regional smallholder associations, so do consider joining your local one. Getting to know more experienced smallholders gives you friendly contacts whom you can chat to, and ask questions about matters that you haven't been able to resolve, such as where to get hold of a few wooden pallets, where to buy small hay bales, the best source for fencing stakes, which local large animal vets can be recommended in the area, or why certain poultry houses are more practical than others.

Cockerels and hens.

3 Land and Accommodation

I am in no way spiritual, with no religion or paganist understanding of the equinox, the mysteries of the moon, or the superstitions that surround the hawthorn and the holly. But, like probably every human that has ever been, I still find huge solace in nature. The whoosh of the sparrowhawk and the subsequent exiting flurry of blue tits and sparrows, the bursting forth of lime-green oak filigree in spring, the glistening clumps of frogspawn spattered with promise in the field ditches, are all rich with relief and reassurance that the world is turning as it should.

A DUTY OF CARE

The life of the smallholder is full of these moments if you just stand still and absorb them. Smallholders actively work with the seasons, not against them, and we have opportunities to enhance rather than damage the earth, and to enjoy what it offers. This is a daily privilege and explains why the smallholder life never fails to attract new blood. The land that we choose to work on, the land that beguiles us into its embrace, is our most precious resource: we can decide how to work with it and enhance it, or we can denude it, and although it is mightily forgiving, we have a duty of care to the land we stand on and all that lives on and in it.

CHOOSING A LOCATION

'But where should my smallholding be?' we ask ourselves, as if there were one perfect map reference destined for each of us. Luckily there is no single smallholding holy grail waiting for you to unearth it. Instead there is a range of things you should consider, from the must-haves to the ideals, and that list will be different for each of us. Just like any property, if you are contemplating moving to a new location, you might need to consider suitable access to employment and schools, but there are additional things to take into account when hunting for a home, which are as much about the patch of land as about the number of bedrooms.

Budget Constraints

Your biggest constraint is likely to be budget; children can change schools and adults seek new employment, and being flexible about location has an impact on what your money can buy. But be careful what you wish for. Moving to the Scottish Highlands for an affordable croft in glorious surroundings will be the answer to the prayers of some, but not all, so ask yourself the hard questions and be honest about your emotions, and your physical abilities and limits. For example, are you happy to be snowed in for weeks of the year, not seeing another soul and being forced to drag feed daily to your livestock on a cobbled-together sled? Will it worry you that friends and family are only prepared to visit in the summer, when you are at your busiest and most productive? Will the fact that it's 10 miles to the nearest shop and petrol station put you off buying, also that the abattoir is six hours drive away, and there is no plumber willing to come out so you have to learn the trade yourself?

It is impossible to give a figure for the budget you'll need. You'll have to work out what you can afford to spend, and finance monthly. Researching what's available and at what price will start to build a picture

Gateway beauty.

of costs in various locations. In some areas places with land rarely come on the market, and when they do they are extremely pricey, so don't waste your time yearning for the impossible. The further you get away from cities the more options there are, but if your income relies on city-based working your attitude to commuting is a crucial consideration. Home-based working is growing all the time, even in sectors that previously thought this was an impossibility, and this is seriously beneficial to the smallholder who wants or needs to continue their existing employment.

In 2022 the cost of agricultural land in the UK ranged between £7,500 and £10,000 per acre. However, if you buy a property and hope to purchase some adjoining land separately, that cost is likely to be multiplied many times; if it's the perfect plot for you, the seller can ask whatever they like.

Quality of the Land

You don't need Grade 1 agricultural land (of excellent quality for growing the most challenging crops such as salads, fruit and vegetables, at high yields) if you want to keep a few hens and sheep. Grade 5 permanent pasture (of very poor quality suited to grazing) would do fine, and you can always create raised beds to enhance veg growing prospects. If a market garden or edible flower enterprise is your intention, you'll need to be fussier about the soil quality than if your ambition is a herd of goats. Trying to plant acres of fruit and veg on stony ground is not good either for your temper, tools or cropping rates. Having heavy clay means we resign ourselves to overwintering cattle in sheds for five months of the year; if this is something you don't want to do (and who could blame you – it's a lot of work), look for land that is free draining in areas with low rainfall, consists of limestone pavement, sandy soils or is heavy on the granite.

Appreciating the slope of the land and which way it faces will affect its productivity, for the same reason that everyone hopes for a south-facing garden to capture the most sun. If the plot is many metres above sea-level in the beautiful uplands, your surroundings will be glorious but the climate and terrain can be harsh, and you, your livestock and your crops need to be tough.

The acidity or alkalinity of the soil will affect what you can grow and how you manage your land, but an acid soil that is not rich in the rye grass much loved by commercial farmers may provide an abundance of wildflowers and other grazeable plants.

Rainfall, Water, Drainage and Sewerage

It is not difficult to research the typical annual rainfall of any particular area, although things do seem to be shifting from past norms quite quickly as we deal with the effects of climate change. Our patch of Devon is green and lush precisely because of the heavy rainfall, so just because your favourite holiday spot is gloriously sunny in August and requires nothing more than flipflops, shorts and a vest doesn't mean you won't have to invest in waterproofs and waders if you live there year-round. As obvious as the impact of heavy rainfall sounds, I've lost count of the number of people who move to the South-West, Cumbria, Wales or the west coast of Scotland without research, and are then surprised by the amount of rain they have to deal with.

Water supply is critical for a smallholder; expect water to be metered unless it is spring fed or comes from a bore hole, but whatever the supply, do ask about its quirks and any limitations before you buy. Boggy patches of land may be marvellous encouragers of wildlife, or simply a result of damaged and neglected drainage, and may scare you off unnecessarily. Resurrecting land suffering from broken drainage systems by replacing pipes choked solid with roots or crumbled by age and decay has a startling effect, turning ground previously impassable during wet seasons into good grazing. The same is true of clearing out ditches that have been trampled into non-existence by livestock over the years.

As for sewerage, we've lived for years with a septic tank and it has been no trouble at all; I think we've had it emptied three times in seventeen years. So don't worry if you don't have mains sewerage, and you will benefit from lower water bills.

PUBLIC RIGHTS OF WAY

If there are public rights of way such as footpaths or bridleways across the land you are considering purchasing, be clear what impact this may have on your smallholding.

- You must avoid putting obstructions on or across the route.
- You must make sure vegetation does not encroach on to the route from the sides or above.
- You must not cultivate (plough) footpaths or bridleways that follow a field edge. The minimum width you need to keep undisturbed is 1.5m for a field-edge footpath and 3m for a field-edge bridleway.
- You must avoid cultivating a cross-field footpath or bridleway. If you have to cultivate, make sure the footpath or bridleway:
  - remains apparent on the ground to at least the minimum width of 1m for a footpath or 2m for a bridleway, and is not obstructed by crops
  - is restored to at least the minimum width so that it is reasonably convenient to use within fourteen days of first being cultivated for that crop
- Where a stile or gate on a public right of way is your responsibility, you must maintain it so it is safe and reasonably easy to use.
- You must not put dairy bulls over the age of ten months in fields containing a public right of way. Bulls over ten months of any other breed must be accompanied by cows or heifers when in fields with public access. Warning notices relating to a bull should only be displayed when it is actually present in a field.
- Horses may be kept loose in fields crossed by public rights of way, as long as they are known to be not dangerous.

Passing Trade

If you are intending to sell surplus eggs, veg, fruit, honey, meat and more to the public, being tucked away in a forgotten hollow or up a mountain, or anywhere truly off the beaten track, is not ideal if passing trade is your main market. On the other hand, such a location may be the perfect draw for campers and holidaymakers eager to buy the ingredients for a breakfast fry-up.

HOW MUCH LAND DOES A SMALLHOLDER NEED?

If your dream of smallholding is having enough space for a few ducks and hens, a couple of dairy goats, a vegetable garden and a small orchard, an acre may be quite adequate. But if keeping a herd of cattle is the key driver for your smallholding, realising this ambition clearly won't be possible on an acre, so size really will matter.

ACRES AND HECTARES

Land size is as often given in acres as it is in hectares. An acre is 70 × 70yd, or 63.5 × 63.5m (4,047sq m). A hectare is 10,000sq m (100 × 100m) and equates to 2.47 acres. To visualise an acre, a football pitch is about 1.5 acres, and if you include the ground usually surrounding the pitch in a stadium, it is closer to 2 acres in total.

The quality of the sward has an impact on the amount of acreage you need; living in the Midlands we could keep twice as many sheep to the acre as we now can in Devon, the grass coverage being thicker and denser and more productive.

If you are fit and not as ancient as me, you may wish to start small but have room to expand as you become more experienced and intrepid. Moving home is such an expense that you may be better off stretching your finances initially (still within a manageable range, of course – too large a financial burden can upset all your plans) to allow you to develop your interests over the long term.

The chocolate-box cottage with a 5-acre paddock is probably one of the most expensive and most desired property options. Instead, consider something that has more land than you think you'll ever need, and a house that may not be prettily swathed in wisteria and yellow roses. Land that is surplus to your initial requirements can earn you an income until you are ready to put it to use by renting it out.

There are as many different designs of smallholding as there are smallholders, but the diagrams overleaf give you a sense of the possibilities on varying sizes of plot. But don't be fooled into thinking these concepts are definitive – you don't have to mimic these ideas to be a 'real' smallholder. Make your own plans, create your own dreams, mix it up and do the things that make you happy.

WHAT MIGHT YOU ACHIEVE ON YOUR ACREAGE?

What you can achieve on your acreage of course depends on whether it is made up of rock, bog, woodland or thickly covered pasture, and what your particular ambitions are. For example, you may wish to dedicate all your space to breeding alpacas for their fibre, and I haven't accounted for that.

Possible Activities on a Certain Acreage

Acreage	Potential activities
Half an acre	At this size you can have a highly productive vegetable and fruit garden, a handful of chickens and ducks, bees, meat rabbits and quail. Hedges can provide foraging opportunities.
One acre	With an additional half acre, you can have the above plus a couple of weaners to raise for pork, and a couple of lambs to raise for meat, or a few geese. With a smaller veg area and plenty of hedges to cut for browse, you could keep a couple of milking goats instead.
Five acres	This size gives you plenty of options. You can add into the mix an orchard, geese, and a small flock of breeding sheep. Dairy (or fibre) and meat goats are possible, or you could keep a couple of cows. There might be woodland for firewood, and enough space for a breeding sow or two. Laying up some grass to make hay is possible. But you can't do all these at the same time.
Ten acres	With 10 acres you can have all of the above, rather than picking and choosing. You can produce enough hay to feed your livestock over winter. Your grass management options improve, allowing areas to rest and regenerate, and you can initiate new projects over time (a vineyard or cider orchard perhaps).
Fifty acres	This size constitutes a small farm: you can have larger numbers of livestock of all kinds, and can produce arable crops for feed and bedding if desired and if land quality and access to suitable equipment allows.

It is critical not to overstock your acreage with livestock, as this will be to the detriment of both the animals and the soil. *See* the later section for indicative stocking rates.

STARTING FROM SCRATCH OR READY-MADE?

Finding a ready-made smallholding that fits your precise dream list is unlikely – and anyway, having an overlong list of 'must-haves' may mean losing out on the property that 'might have been perfect if only I hadn't been obsessed by the inadequate galley kitchen'. Having initially discarded a property precisely because the kitchen was tiny and incredibly dark, I drove past a few weeks later and realised what promise the whole package actually had – and soon after it became our first smallholding. A few years on we knocked the kitchen through to the dining room that we never used – and the house went from ugly duckling to swan.

The deluge of property programmes on television serves to highlight how a single-minded hunt for a place with the perfect this or that is likely to end in disappointment, and how compromise has to be squarely embraced. The one thing you can't change is the location, so focus on potential, rather than fretting over the elements you can improve over time. See an avocado bathroom suite and a crumbling

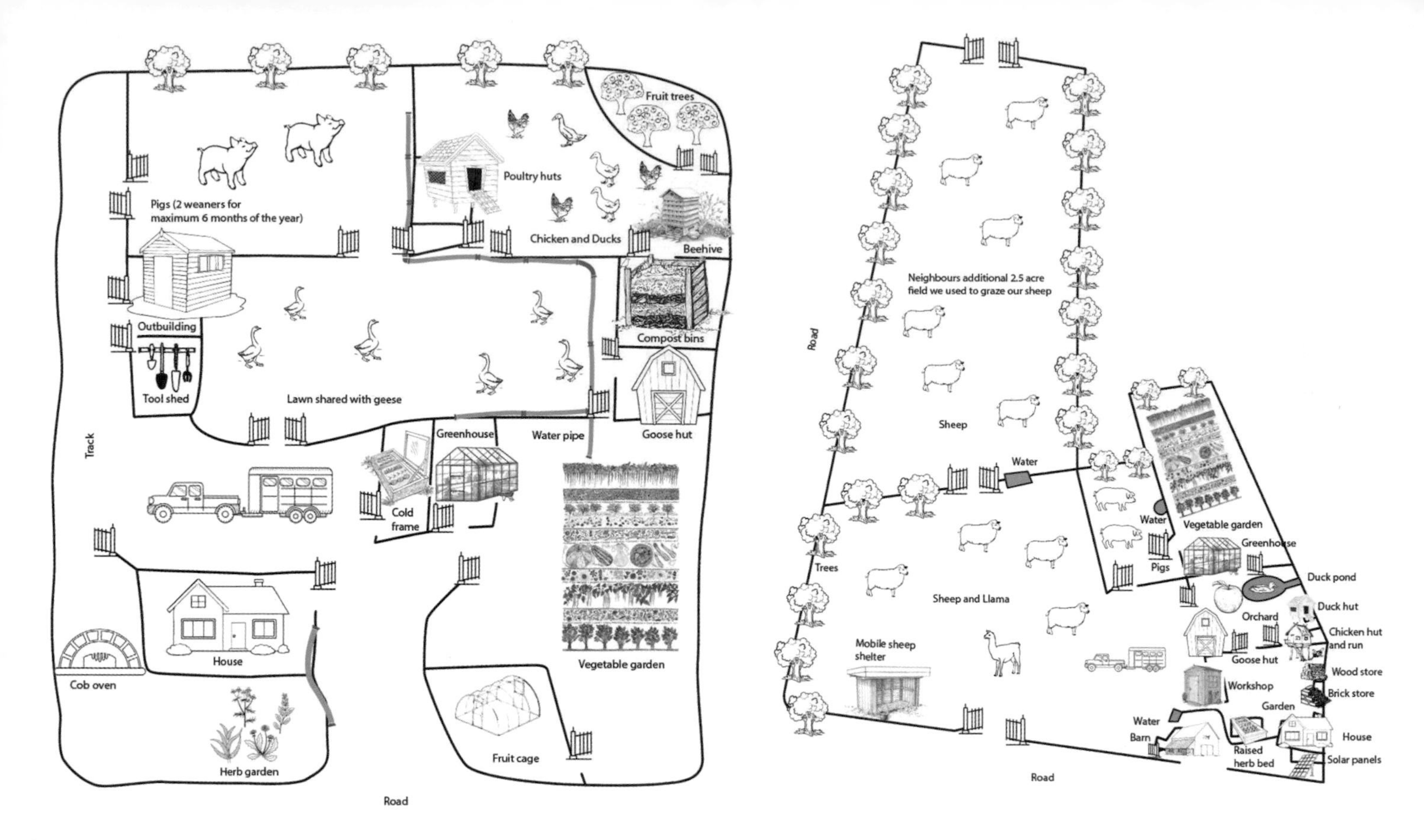

One acre.

The author's first smallholding (3.25 acres)

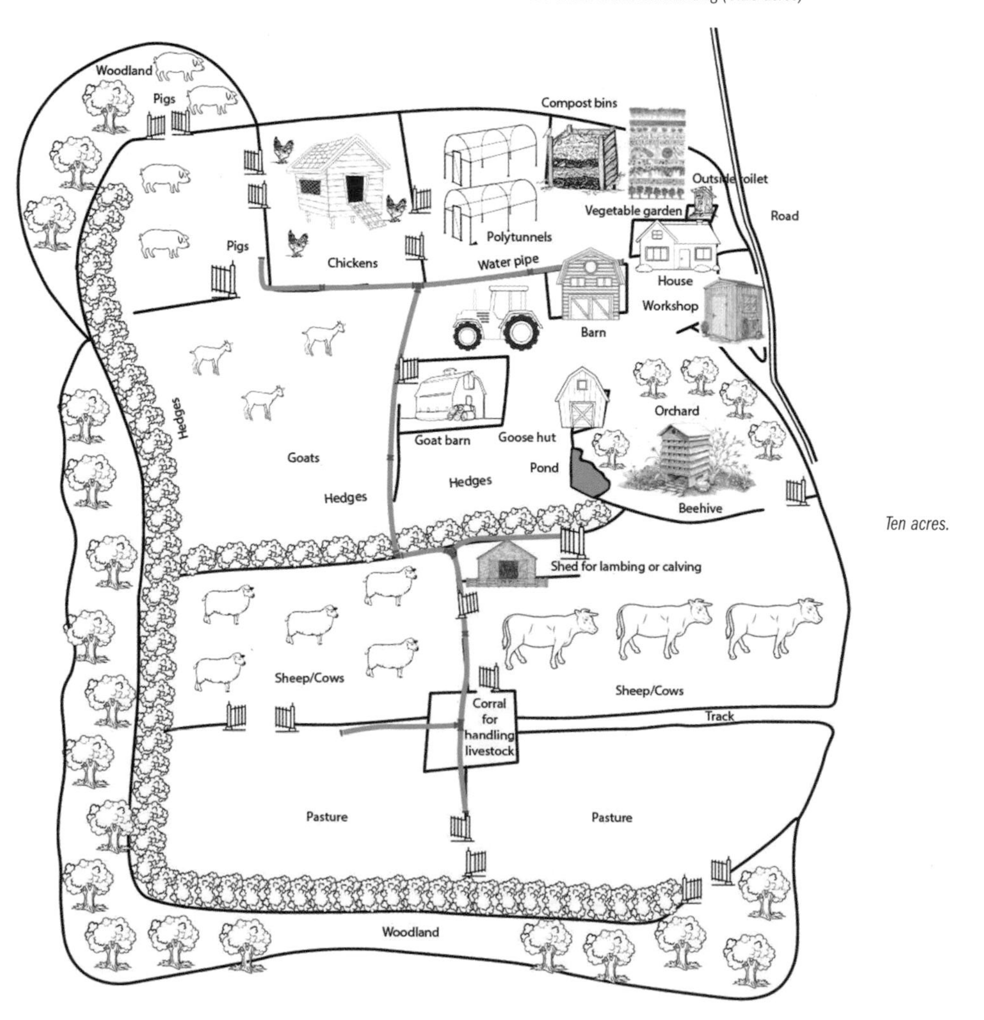

Ten acres.

Derelict cowshed.

Restored cowshed.

pigsty as projects, not insurmountable defects. When we moved to Devon the farm was in a truly dreadful state, but a decade on, and the main failings had been addressed. We have usable barns with roofs and walls, the land is fenced, the hedges, ditches, drains and woodland restored. The situation of the farm, down a lane with grass growing in the middle of the road, surrounded by its own land, offered space, quiet, air to breathe, and a life seated in natural beauty.

The house-related things I'd find challenging to live without include peace and quiet; an outside or easily accessed downstairs toilet; a utility space for boots, wet weather gear, livestock medication, washing machine and dogs; a kitchen that's more than a cubby hole; an office space; broadband; a sofa to snooze on; a shower and a comfortable bed. And that's pretty much it.

PLANS FOR YOUR SMALLHOLDING

People who are really clear about the shape of their smallholding ambitions are quite rare; most have a fancy for a few chickens, perhaps a couple of pigs, a veg plot and a few goats or sheep if the land allows, but are open to other possibilities. Unless you have absolute requirements (a vegetable garden that feeds the whole family year-round, which might dictate that living on a rocky mountain is not going to work), you may be happy to let the chosen holding shape your next steps. You may not be particular about whether to keep sheep or goats, and a holding with a lot of scrub and browse may direct you more towards goats. An exposed plot by the coast may mean that a plum orchard is not going to be practical, but you can grow a sheltering hedge and have a few dwarf fruit trees against the south side of the house. The plot shapes your ambitions and generates ideas, and there isn't time or inclination to fret about the limitations.

If you do have very specific plans, make sure the place you choose can accommodate them. It sounds absurdly obvious to say that, but the attractiveness of a holding can grasp you firmly by the heart. In

Jersey house cow and calf next to a pig paddock.

the excitement of finding an inspiring new home it is possible to forget that your primary aim was to have cows and a micro dairy to explore a future as a small-scale cheesemaker, and overlook the truth that the land is entirely unsuited to the few Jersey milkers that have populated your dreams.

Outbuildings, Sheds, Barns, Lean-Tos and Shelters

The smallholder who doesn't have any sort of outbuilding faces a serious challenge – and to be frank, the more you have, the more helpful it is. If you take on somewhere with clusters of old buildings, please don't spend the first year razing them to the ground. Once you start smallholding, you'll probably want to use any and every building's footprint and possibly its structure to repair, rebuild and restore. Flattening everything and starting from scratch may cause planning headaches, and on a practical note, if previous owners have been canny, the siting of old buildings is usually the best location for them in terms of drainage, access, prevailing winds and ease of movement from place to place. There are exceptions, of course, where the siting of an ugly building in front of the kitchen window may be an affront to aesthetics and practicality.

You'll need somewhere dry to store hay, straw, animal feed and equipment. You might want a small dairying space; all goats will need a shelter, and like us you may prefer to lamb, kid, calve and farrow inside. A space to quarantine poorly animals is useful, or having a shed with adequate human headroom to house your poultry (this is a boon when mucking out, and also when overwintering poultry inside because of the increasing likelihood of seasonal avian flu lockdowns). A simple workshop where repairs can be done and tools kept is all but a necessity, plus somewhere to store your produce, whether in freezers, jars or loose on shelves.

If you plan to build a barn, space allowing, make it bigger than you think you'll need. And don't butt it up against a corner boundary, because in a few years you may be grateful that you can add on lean-tos to more than one side.

Stocking Rates

Every smallholder wants to know what numbers of livestock a plot of land can sustain. The answer is dependent on many factors, including the sizes and breeds of the animals you plan to keep, the quality of the land, whether you expect to make your own forage, and whether you feed concentrates or not. Suggestions are given in the table below, and they are deliberately generous on land allowances, ranging from larger to smaller breeds, and assuming that you buy in forage for the winter months and are therefore not using any of that land to make hay.

Always underestimate stocking rates; you can add to your numbers if the land proves more productive than you anticipate, and if you have breeding animals, numbers will quickly increase. And remember that you can't have all of these options per acre: if you have 1 acre and five ewes and lambs, that's your allocation completely used up.

Stocking Rates

Species	Numbers per Acre
Chickens	50–100
Ducks	24–40 large breed ducks
Geese	8–16 adults
Sheep	3–5 ewes and their lambs
Goats	2–3 goats and their kids
Pigs	3–4 sows with litters 24 weaners (7–18kg per weaner) 11 growers (18–35kg per grower) 7 porkers/baconers (35–90kg)
Cattle	<0.5 cow (2 acres for a cow and its calf)

TIDYING UP NATURE

There are plenty of places for tidiness on a smallholding: the neat heap of holding registers, the veterinary medicine book and the folder for movement licences; the bucket of baler twine ready for use; the garden tools hung from hooks; the well stacked bales of straw and hay. But please avoid being an inveterate tidier of nature. In our courses we are asked all the time how ancient ridge and furrow land can be flattened, how hillocks can be smoothed away, how nettle patches in an untended corner can be removed forever, or how a stream can be straightened.

Don't be too tidy.

The answer is to leave well alone, or even to encourage nature to make it messier yet. Let gorse and brambles spill through your fencing; they will create a much enjoyed larder for yourself and wildlife, and will stop livestock pushing against the fence. Allow what *you* might consider weeds to flourish as hosts to insects where they don't interfere with you. Don't spend your life on a sit-down mower removing every wildflower or broad-leaf plant – this is a smallholding, not a bowling green. If you must tidy, save it for marshalling your chutneys in lines on their shelves, keeping your wheelbarrows in rows, or parking any tractor implements in a satisfactorily neat formation.

This isn't to say that every tree, shrub, copse and rill should be left as it is. Drainage ditches need maintaining, ponds can be created, hedges benefit from trimming and laying just as roses do from pruning, new trees and hedges should be planted, and sour land enlivened. There is a lifetime of reading and study available to you on restorative agriculture, improving soil structure and biodiversity, regenerative grazing practices and sustainable livestock production. The principles apply equally to a 500-acre farm and a 3-acre smallholding. Pockets of unmown sward will be rich in insects that will encourage bats and birds, voles, slow worms and butterflies, and if you're lucky your compost heap will contain grass snakes.

BUYING, RENTING, BORROWING, COMMUNITY SHARING AND TENANCIES

Although most smallholders hope to buy a smallholding – a patch of land and a house to live in – as one tidy package, this may not be possible. But there are other options, such as renting land not too far from home, or using land that another smallholder can no longer manage, for reasons of age, health, work or family demands. There are community land shares and a very few council tenancies, and you may find that housing and a piece of land is part of an employment opportunity.

Hunt out local allotments, talk to neighbours with garden to spare, or investigate small plots of land that are looking unloved and ask the owner if you can tend them (the Land Registry provides details on land ownership if asking locally fails to provide that information). I've had numerous conversations with determined cash-poor smallholders-to-be who have spoken to all the farmers and landowners in their vicinity, finally resulting in an agreement to use a piece of land. Many people rent a field at some distance from home in order to get started.

Agricultural Ties

An Agricultural Occupancy Condition (AOC), more commonly known as an agricultural tie, is a covenant placed on a property by the local authority restricting its occupancy to workers who are solely or mainly actively involved in agriculture as their main occupation. Ties are used by planning departments to stop people building houses claiming that they are needed to house an agricultural worker, and then selling them on the open market. It may be possible to have the tie lifted, but this is by no means assured.

To try to have the AOC removed the owner is required to put the property on the open market for anything up to two years at its genuine value, which is likely to be around 65–75 per cent of the value it would have without the AOC. If it can be shown that despite their efforts there are no buyers who meet the AOC criteria interested in purchasing with the AOC in place, the owner can apply to their local planning authority to have it lifted.

An alternative is to apply for a Certificate of Lawful Use (CLEUD), where the local planning authority certifies that an existing building use is lawful. It has to be proven that the people living in the property for at least ten years have already been in contravention of the AOC rules for that period.

HEART AND HEAD

You may have sensibly drawn up a checklist to guide your smallholding search, but if you are anything like me, you'll stand at the entrance gate, blind to the 'Danger, do not enter' signs plastered over the barns, gaze at a long, whitewashed building with yellow doors and window frames, and say: 'This is it!' – and hopefully it will be.

4 The Legal Aspects of Smallholding

Smallholding is substantially an outdoor activity, plus plenty of opportunity for DIY, workshop, kitchen and craft creativity, and I'm not sure whether it's a boon or a bind to reveal that there is just enough paperwork involved to satisfy your inner bureaucrat.

Let's start with a quick word of reassurance that implementing the legal aspects of smallholding should be neither time consuming nor overly complex. There are things to learn about, and requirements that must not be shirked, and probably this will be a new experience for the first-time livestock keeper (it's the livestock aspect that has the greatest demands), but the reality is that this should takes minutes of your time over a year, not days. The difficult and time-consuming part is building your knowledge of the rules: once these have been grasped, smallholding, not the paperwork, can have your full attention.

The first concept to get your head around is that whether you are keeping a couple of pigs as pets or a thousand sheep, the rules apply equally to all. Livestock are food-producing animals even if they aren't intended for the plate, and there is no distinction with regard to your legal obligations between commercial and hobby keepers, and no exemptions for even the most pocket-sized of smallholdings.

The rules do change, and it is the responsibility of each livestock keeper to keep up to date with any new requirements (in the UK via www.gov.uk). The contact details for the various bodies are in the further information sources in the appendices.

British Lop piglets.

There is some frustration that the rules are not the same for pigs, sheep, goats or cattle, so you cannot assume that understanding the rules for one species means they are the same for another: they are not.

STEP ONE: REGISTER YOUR HOLDING

Registering your land, even if it is a large back garden where you keep a couple of pet pigs, means acquiring a County Parish Holding (CPH) number for the land where the livestock will be kept; there is no charge for this. Its main purpose is to identify and trace the location of livestock, and it is a unique nine-digit number: the first two digits relate to the county, the next three to the parish, and the last four are unique to the keeper.

To apply for your CPH number contact the Rural Payments Agency in England, Rural Payments Wales in Wales, and Rural Payments and Services in Scotland (there are no CPH numbers required in Ireland or Northern Ireland). If you intend to keep livestock on someone else's land you still need a CPH. You'll need the National Grid field numbers for the land intended for livestock, located using an online tool at https://magic.defra.gov.uk/. If you have land at various sites that fall within a 10-mile radius of the main livestock area, these can be included in a single CPH. A permanent CPH covers land used for more than a year, a temporary CPH (tCPH) covers the use of land up to a year.

Once you have your CPH you can move livestock to your holding under a general licence (usually known as a movement licence; more on that below).

LIVESTOCK NEEDING A CPH NUMBER

If you keep one or more of the following, you need a CPH number:

- cattle (including bison and buffalo)
- deer
- sheep
- goats
- pigs
- poultry (more than fifty birds).

You do not need a CPH to keep camelids (alpacas, llamas), donkeys or horses.

STEP TWO: OBTAINING HERD AND FLOCK MARKS

The next step is to obtain herd or flock marks for the livestock you intend to keep; this is also a free service. Contact the local Animal and Plant Health Agency office (APHA) within thirty days of livestock moving on to the land. You will need these when ordering ear tags. The APHA can provide herd numbers for camelids if a CPH is already registered, but it is not necessary.

If you have fifty or more poultry (collectively hens, geese, ducks, quails, turkeys numbering fifty or more) you also have to register your flock within one month of their arrival; search for poultry registration at www.gov.uk. It's advisable to register voluntarily even if you have fewer than fifty birds as you will be alerted to any disease outbreaks such as avian influenza that require taking biosecurity measures.

THIRD STEP: IDENTIFYING INDIVIDUAL ANIMALS

Individual animals are identified by means of ear tags and other identifiers, described in the following sections.

Sheep

Sheep ear tags have the flock number of the holding on which they were born, plus a unique number for that sheep. Sheep keep the same tags in for life, so the first sheep brought on to your holding will have tags that will not have your flock mark and you do not change these tags. The first lambs born on your holding will be the first to be ear tagged with your own flock mark. Sheep born on your holding must be identified within six months of birth for intensively reared stock (defined as those housed overnight), or nine months of birth for extensively reared stock (not housed overnight), or when they leave your holding, whichever is first.

Ear tags and applicators.

You will need to apply two tags, one in each ear, to sheep kept or sold for breeding, and any that will be kept over twelve months. One of the identifiers must be EID (electronic identification) containing an electronic chip with the same information as is written visibly on the tag); all sheep EID ear tags are yellow. The second is called a visual tag and can be any other colour you fancy that the

Ear tags in Herdwick sheep.

manufacturer produces apart from yellow, black or red; all numbers on both tags must match. It is best practice to put the yellow EID tag in the sheep's left ear.

For sheep intended for slaughter before twelve months of age you can apply a single yellow EID slaughter tag, or full double tagging. If you are not sure whether an animal is intended for breeding or slaughter at the time of tagging, you should apply two identifiers.

REPLACEMENT TAGS

If you discover an identifier has been lost or damaged, you must replace it within twenty-eight days of noticing. If you have to remove a tag from an animal because of an ear infection, you must replace it with an identical replica tag as soon as the infection clears up. Replacement ear tags applied on a holding other than the one where the animal was born must be red, or if you know the numbers that were on the missing tag you can use replicas, an exact replacement of the original tag.

Goats

The rules for goats are similar to sheep with the following differences:

- You can choose whether or not one of the two identifiers is a yellow EID tag – you may opt for two visual tags.
- For goats intended for slaughter before twelve months of age, a single identifier is adequate and it need not be an EID.

The majority of goats are identified using double ear tags (one tag in each ear), but you can use one ear tag plus other options.

On a practical note, goat ears are less robust than those of a sheep, and a combination of delicate ears, a desire to stick their heads in the most inconvenient of places, and with some breeds having ears of distinctive length, it's important to use tags that minimise the risk of getting caught and tearing the ear. Two-part tags can be purchased, or if you use single-piece tags, do cut the hinge part with a pair of snips so that the two parts swing and don't form a trap hazard. A pastern band is a tamperproof ankle bracelet and is ideal for animals with damaged or sensitive ears, and doesn't require an applicator.

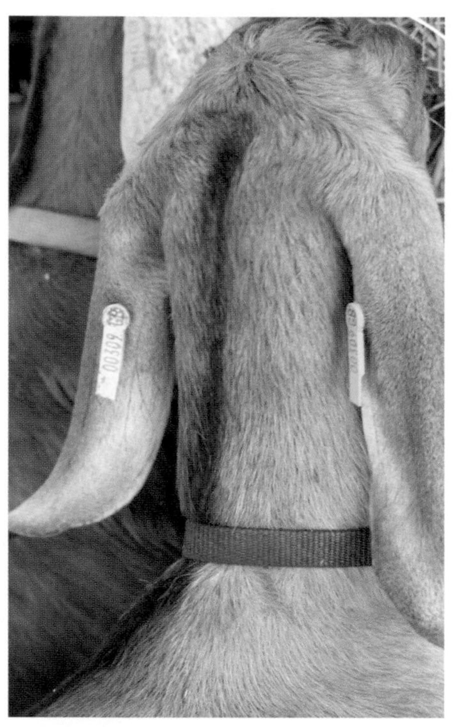

Goat ear-tag placement.

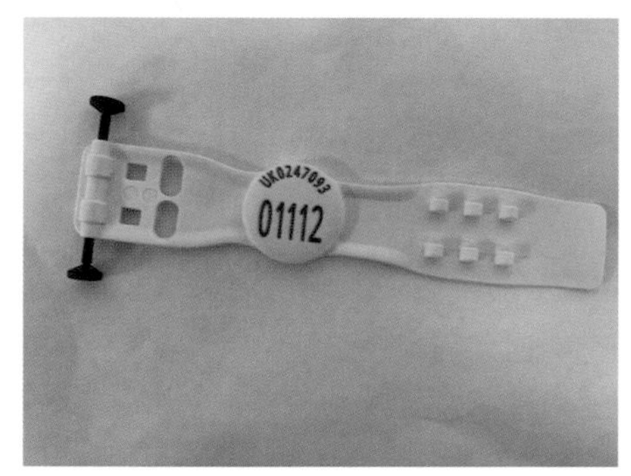

Pastern band (Dalton Tags).

OPTIONS FOR IDENTIFYING SHEEP AND GOATS

First identifier	Second identifier
Sheep and goats	
Yellow EID ear tag	Visual ear tag
Yellow EID ear tag	Tattoo
Yellow EID ear tag	Pastern (leg) band
EID bolus	Black ear tag or pastern
Yellow EID pastern (leg) band	Visual ear tag
Goats only	
Visual ear tag	Visual ear tag
Visual ear tag	Tattoo
Visual ear tag	Pastern (leg) band
EID injectable chip (in groin)	Black ear tag

Pigs

All pigs over twelve months need to be identified with a herd mark. However, pigs going to slaughter must have a tag/permanent mark from the holding they are travelling from, *not* their place of birth if that is different. Unlike sheep and goats that are tagged on their holding of birth, if you are buying weaners intended for meat, you do not want the breeder tagging them with their own herd number as you will need to tag the pigs with your ear tag/herd number before taking them to the abattoir (*see* Chapter 9 on how and when to do this).

Pigs going to slaughter must have an identifier capable of surviving the processing of the carcase – this is usually a metal ear tag, or a permanent tattoo on one of the pig's ears, or two slap marks applied to both of the pig's front shoulders using permanent

Pig ear tags and applicator.

ink. Neither slap marks nor tattoos are appropriate for dark-coloured pigs; in my experience a metal ear tag is generally the simplest option.

Movements between holdings in the case of pigs under a year should use a temporary stock-spray paint mark, such as a blue cross, yellow dot or white stripe that can be noted as such on the movement licence: this mark must last until the pig reaches its destination. If you are buying pigs over twelve months old, for example as breeding stock, they should be tagged with the breeder's herd number.

Cattle

All cattle must have an approved ear tag in each ear, showing the same unique number. The primary tag is always yellow and of specified size (large and easy to read), and the secondary tag can be the same size or of different design and includes identical information. Tags must be fitted at the same time, within twenty days of birth for beef calves, and dairy calves must have at least one of their tags fitted within thirty-six hours of birth. All cattle must be tagged before they leave the holding of birth. You should not purchase more ear tags than you'll use in one year, and any unused ear tags should be stored securely.

All cattle born in or imported into Great Britain need a valid cattle passport (no photo is required) that accompanies the animal for all movements. The breeder must register the birth of each calf online, which generates a cattle passport that arrives by post within a few days. All birth registrations must be made within seven days of tagging, which gives all keepers of beef cattle a maximum of twenty-seven days in which to tag the calf and register the birth. At the time of writing, births and cattle movements need to be registered with the British Cattle Movement Service; it is expected that the Livestock Information Service will take on this role.

Each make and type of ear tag requires its own applicator.

Primary and secondary cattle tags.

MOVING LIVESTOCK

Movement Licences

A licence form must be completed every time you move livestock on or off your premises – whether

Movement Document under the Sheep and Goats (Records, Identification and Movement) (England) Order 2009 (as amended).

Department for Environment Food & Rural Affairs

Please complete in BALLPOINT pen and press firmly but before doing so read the important notes overleaf.

For completion by Keeper at Departure Location

1. Departure Location

Departure CPH

Keeper's name and full postal address of holding of Departure

Postcode

Full postal address of Destination

Postcode

Sheep ☐ Goats ☐ No. of Animals	Individual identification number(s) or For Slaughter Animals - the Mix of Flock/Herd Mark(s)
Write total number of animals moved here	<<<Total Number of Animals moved

Please tick box if:

Moving animals to a Central Point Recording Centre (CPRC) ☐
Moving animals within your business that remain under your day to day care and control ☐
Return after rounding up from common land ☐
Using a supplementary sheet or attaching a list of individual identification numbers (stapled to this form) ☐
Destination Exhibition or Performance with permit (Non CPH) ☐
Moving from/to* an approved isolation unit (*delete whichever does not apply) ☐
Moving animals to a collection centre for immediate slaughter ☐

Food Chain Information (FCI) - for sheep and goat movements to slaughter only **Tick one box**

I have read the FCI statement on the reverse of this form and declare that all animals in this consignment satisfy its conditions that allow them to enter the food chain. ☐

OR

The FCI Statements are not satisfied for all the animals in the consignment and additional information is provided on the reverse of the **pink copy** or on an attached document. ☐

I declare that the above details are correct

Keeper's Signature

Print Name

☎

Email

Name of Owner if Different to Keeper Named Above

Departure Date (DD/MM/YY)

Date of Loading (if different) (DD/MM/YY)

Time First Animal Loaded (HH/MM)

Time of Departure (HH/MM)

Expected Duration of Journey (HH/MM)

For Completion by Transporter

2. Transport Details **Tick box to indicate who is transporting animals**

Departure Keeper ☐ Receiving Keeper ☐ Haulier ☐

Vehicle Registration

Haulage Company

Transporters Authorisation Number (Where journey is over 65km)

Haulier's Signature

Print Name

☎

For Completion by Keeper at Receiving Location

3. Receiving Location (All sections must be completed)

Destination CPH or Slaughterhouse No.

☐ **Tick box if location is a dedicated slaughter market**

Number Received

☐ Tick box if keeper has **NOT** changed. If this is the case, boxes below are optional

I declare that the above details are correct

Keeper's Signature

Print Name

LIS-1 03/22

Arrival Date (DD/MM/YY)

Date of Unloading (If different)

Time Last Animal Unloaded (HH/MM)

☎

Email

Please return to: Livestock Information Service (LIS), PO Box 6299, Milton Keynes MK10 1ZQ
For further information visit www.livestockinformation.org.uk

090965 (PROOF 2) Front Parts 1 - 4 Black

Movement licence for sheep and goats.

to slaughter, or buying in or selling stock, taking them to a show, or to a holding with a different CPH; this is a free service. The place of departure is responsible for generating the movement licence, and the receiving destination is responsible for confirming that the movement has taken place. Increasingly these forms are completed online, and the system is likely to become entirely digital in time, although a paper-based option for sheep and goats is still available. An online Livestock Information Service (initially for sheep and goats, progressing to include pigs, cattle and deer) was launched in England in 2022, with similar services in development by the Scottish, Welsh and Northern Ireland devolved administrations. You do not need a movement licence if the movement is to or from:

- a vet for emergency treatment
- land with the same CPH number
- land bordering your holding with a different CPH number, where you send or receive animals from frequently, on foot and without contact with other livestock
- common land bordering your holding that you've registered as linked to your holding.

Food Chain Information

When taking livestock to the abattoir you will need to complete the relevant Food Chain Information section included on the movement licence, which confirms (or otherwise) that the animals being taken to slaughter have passed the withdrawal period of any administered medication, and are fit to enter the human food chain. A number of abattoirs have an additional livestock delivery form that they require to be completed when delivering livestock, particularly if the majority of their business is serving the retail and wholesale trade rather than private kill (for more on abattoirs *see* Chapter 14).

Movement Standstill

Movement standstill is designed to slow down the spread of disease. Whenever cattle, sheep, goats or pigs are moved on to a holding, no cattle, sheep or goats may move off for a period of six days. For example, if you move a goat on to your holding, no goat (including the one just brought on), nor any cow, sheep or pig, can move off until the six-day standstill has elapsed; animals can be moved on the seventh day.

The movement of pigs generates a longer standstill: where pigs have moved on to a holding, that pig or existing pigs must remain under standstill for twenty days, and can only move off the holding on the twenty-first day (other species can move off on the seventh day). In Scotland the standstill period is thirteen days for cattle, sheep and goats, but the same twenty-day standstill for pigs.

There are some exemptions to movement standstill. You can move livestock during the standstill period if moving them direct to the abattoir. If you have commoner grazing rights, animals may move between common grazing and its associated in-bye land without triggering a standstill period. Breeding animals of either gender do not trigger a standstill on premises to which they are moved for breeding, provided they are isolated for six days (twenty days for pigs) in an APHA-approved isolation facility (*see* box) before they leave the premises of origin.

Livestock may attend shows without having to observe a movement standstill period, provided that they have been isolated from all other non-show animals on the holding in an approved isolation facility for six days (twenty days for pigs) before attending the show. On their return, the show animals must be kept in an approved isolation facility for six days (twenty days for pigs) or the standstill will be imposed; they can, however, go to other shows within that period. You will still need to complete a movement licence for all standstill-exempt movements.

APHA APPROVED ISOLATION FACILITY

If you are planning to show livestock regularly, or take them to another holding for breeding, you may find it helpful to set up an isolation unit. The forms for doing so are available from APHA, and the facility will need to be approved by a veterinary inspector. An isolation unit, whether pasture or building, must be dedicated for livestock isolation purposes and be physically separate (a minimum of 3 metres) from any other livestock buildings or pastures. Manure and effluent from isolation facilities must not come into contact with other livestock, and dedicated protective clothing and footbaths must be used in the isolation facility.

HOLDING REGISTER

To protect the health of your livestock and to make it easier for APHA and other relevant bodies to trace your animals, you must keep a register of the animals on your holding. This includes information about your holding, tags and replacement tags in use, and the movements on and off your holding. Electronic versions are freely available to download from www.gov.uk (search for 'sample sheep and goat holding register'). In addition to movements, the register should include all livestock births and deaths.

How Long do I have to Keep Records?

Holding register	Keep for ten years from the end of the calendar year in which the last entry was made.
Veterinary medicine records	Keep records of treatment given to animals, and of animal mortality for at least three years.
Paper movement licences	Keep a copy for three years from the date the animal arrives on your holding.

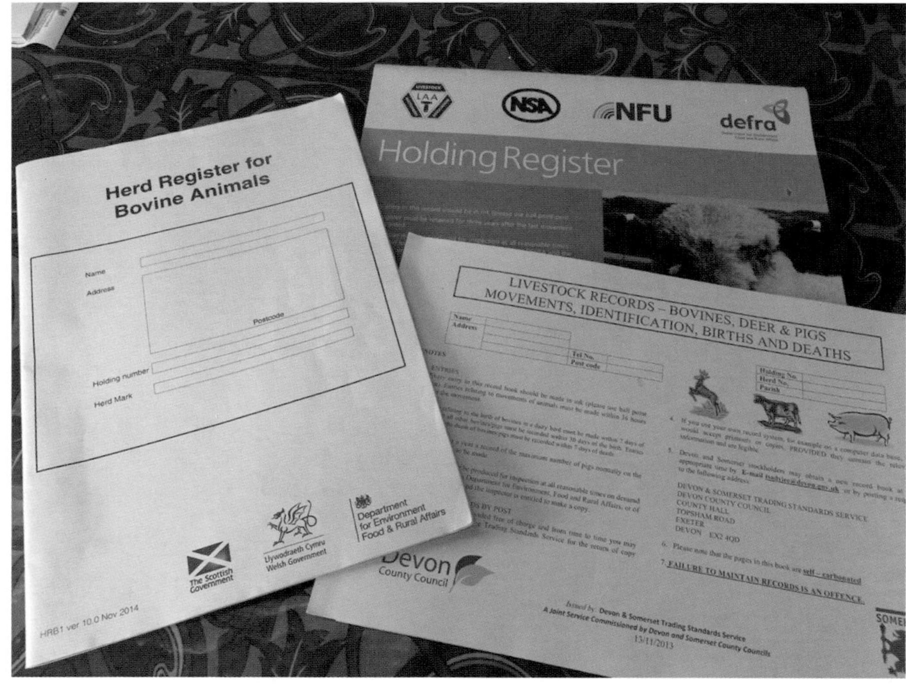

Examples of holding registers.

VETERINARY MEDICINE RECORD

It is a legal requirement to keep a record of all veterinary medicines administered to food-producing animals, including those administered by your vet or given in feed, even if you never intend to eat your livestock or use their milk or eggs. The record must show the following: the name of the medicine used, the supplier, the date of purchase, the date of administering the medicine (and the end date if it is over a period), the quantity of medicine used, the identity of the livestock treated, and the withdrawal period for meat, milk or eggs as appropriate, the batch number of the medicine, and the expiry dates.

The batch number will be written on the packaging, and the withdrawal period will either be written on the packaging or included in the datasheet inside. There are livestock medicine record books available from most agricultural merchants, and electronic versions are freely available to download. It is advisable to complete your medicine records immediately after any treatment when the information is fresh in your mind.

TRANSPORT CERTIFICATE

You must have your competence independently assessed if transporting livestock by road on journeys over 65km in connection with an economic activity. If your livestock is kept purely as a hobby and there is no commercial element involved (you don't sell livestock, meat, milk or dairy products), then this will not be relevant. The assessment will be as follows:

- A theory test for journeys over 65km and up to eight hours.
- A practical assessment of competence including animal handling, and if required, driving skills for journeys over eight hours.

If assessed as competent, you will receive a certificate of competence. The certificate will be specific to your role of either transporter or attendant, the length of the journeys you take, and the species you transport. Certificates of competence are valid for life. You also need to ensure you have access to a suitable trailer to transport your livestock. Contact www.lantra.co.uk for information on training and assessments for transporting livestock.

Veterinary Medicine Record

Devon County Council

The Animals and Animal Products (Examination for Residues and Maximum Residue Limits) Regulations 1997 as amended
The Food Hygiene (England) Regulations 2006
The Welfare of Farmed Animals (England) Regulations 2007
The Veterinary Medicines Regulations 2013

Name:
Address:
Post Code:
Telephone:

Holding Number:
Flock / Herd Number:
Parish:

The Veterinary Medicine Regulations 2013 require the following to be kept by farmers:

- The keeper of a food-producing animal must keep the documentation on the acquisition of a veterinary medicinal product and the records relating to the product for at least **five** years following the administration or other disposal of the product, irrespective of whether or not the animal concerned is no longer in that keeper's possession or has been slaughtered or has died during that period
- The keeper of a food-producing animal must keep proof of purchase of all veterinary medicinal products acquired for the animal (or, if they were not bought, documentary evidence of how they were acquired
- **When a veterinary medicinal product is bought or otherwise acquired for a food producing animal, the keeper must at the time record -**
  1. the name of the product and the batch number;
  2. the date of acquisition;
  3. the quantity acquired; and
  4. the name and address of the supplier.
- **At the time of administration the keeper must record -**
  1. the name of the product;
  2. the date of administration;
  3. the quantity administered;
  4. the withdrawal period; and
  5. the identification of the animals treated.

TRADING STANDARDS SERVICE
DEVON COUNTY COUNCIL
COUNTY HALL
TOPSHAM ROAD
EXETER
EX2 4QH

Email: tsadvice@devon.gov.uk

1/10/13

Veterinary medicine record.

TAKING YOUR PIG, GOAT OR SHEEP FOR A WALK

In England and Wales, permits can be issued to take pet sheep, goats or pigs for walks. The permit disapplies the six-day standstill on return to the home premises. Keepers of pets need to apply in writing to their local APHA office, and include a description and/or map of the route to be approved. The route will need to be risk assessed and approved by a veterinary officer (VO) at the local APHA office. As part of the risk assessment the VO will need to decide whether a visit is needed. APHA may not approve your walking route if it poses a health risk – for example, if it passes close to a livestock market, farm or fast-food restaurant. You must have your licence with you whenever you are walking your pet livestock, and you'll need to renew it every year.

DISBUDDING, DEHORNING, CASTRATING AND TAIL DOCKING

Disbudding

Disbudding is the cauterising and destroying of the horn buds before they can grow into horns. In the UK only a vet can disbud goat kids, and it should be done, if necessary, in the first week of life, ideally at two to three days old. The operation is not without risk, and keeping an entirely horned goat herd is a valid option, although there are inevitably more opportunities for goats to get themselves caught in fencing. As it is inappropriate to mix horned and hornless goats because of the risk of bullying and subsequent injury, this is an issue to consider right at the beginning of keeping goats.

Most dairy herds will have their kids disbudded, as horns can cause serious wounds to udders in particular. Disbudding of calves should take place before calves are two months old and ideally much earlier, in fact as soon as the horn bud can be seen/

Vet Chris Just with disbudded kids.

felt. The procedure should only be carried out with a heating iron, using local anaesthetic; if this is beyond the competence of the smallholder, your vet should carry out the task.

Dehorning

Dehorning – the total removal of a horn of an older animal of any species – should never be seen as a routine procedure or done for cosmetic reasons; it is a significant operation, and both disbudding and dehorning are considered mutilations. If necessary for the welfare of the herd or individual animal it should be carried out by the vet under local anaesthetic in spring or autumn to avoid flies or frosts affecting the wound. Animals should be given appropriate pain relief and the wound protected until the hole has scabbed over.

Sheep are rarely dehorned unless the horn is found to be growing too close to the head, threatening to grow into the skin or the eye, in which case this must be attended to before the horn digs into the animal's face. In young horned sheep the tips of in-growing horns can simply be clipped with foot shears, and these may need redoing as the sheep grows. In older sheep where the horn or horns pose a problem, use a wire saw. Wrap and securely pin the ends of the wire to two handles – thick dowel or short pieces of broom handle work well – leaving

Goat with water pipe taped to its horns to prevent it getting its head stuck in fencing.

Wire saw – in this case used to remove an ingrowing horn scur.

a usable length of 60–70cm of wire. The sheep will need to be restrained, and only saw off as much of the horn as necessary. The person sawing off the horn needs to saw quickly and consistently with the wire pulled taut; the heat this creates cauterises the blood vessels.

Castration

Castration isn't always necessary, but there are husbandry reasons for deciding to do so. Entire, uncastrated males can mate with their dams, sisters, and any other available female from a surprisingly early age, as young as six weeks in the case of precocious bucklings, resulting in unwanted pregnancies. Smallholders may not have the space to separate adolescents from their dams at weaning, as well as separating males from young females.

Castration (also known as ringing, banding or wethering) of ram lambs, bull calves and bucklings must be carried out during the first week of life and by a trained competent person, so for your first time do ask an experienced shepherd, goat or cattle keeper or your vet to show you how it's done. It will not be possible to castrate in the first week of life if one or both testicles have not descended. You can either leave the male entire, or use a burdizzo for bloodless castration by crushing the spermatic cords of youngstock less than two months old (this should only be done by someone competent). After two months of age (three months for sheep) castration is only allowed by a veterinary surgeon, using an anaesthetic.

The rubber castration ring constricts the blood flow, and in three or four weeks the scrotum will have shrivelled up and dropped off. The procedure of ringing is as follows:

- Put a rubber ring on the elastrator pins. For ram lambs and bucklings, hold the animal by the front legs so that the back legs and scrotum dangle down. For bull calves, sit them on their bottom.
- Open the elastrator and place the ring over the scrotum at its base against the body.
- Close the elastrator but do not remove it.
- Make sure both testicles are below the ring before you remove the elastrator (they feel like two firm rods inside the scrotum).
- Release the ring from the elastrator. Check that the nipples are rolled above the ring and are free of it.

Elastrator and castration rings.

- If you have caught only one testicle in the band, remove the band immediately with a sharp knife or blunt-nosed scissors and try again later. We keep a small pointed piece of wood and a sharp penknife in the pot of castration rings – the wood is for inserting between the ring and the animal's skin so that you can cut off the ring without damaging the animal.

Do not castrate calves, kids or lambs until they are a couple of days old and have bonded well with their dam; castration is undoubtedly painful for kids and lambs for anything from a few minutes up to an hour or more, and you don't want to interfere in any way with colostrum intake, suckling and bonding. The process seems to have little or no negative effect on bull calves.

This early castration on goat kids using an elastrator and rubber ring can restrict the internal diameter of the urethra, which as a result doesn't grow to full size as the wether matures, increasing the possibility of obstruction by urinary calculi. This is a particular risk for those intended as long-term pets as they mature; the issue is of less concern for those that go for meat within their first year. This being the case, consider the alternative of later surgical castration by the vet.

Tail Docking

Tail docking is commonly used for downland breeds of sheep and rarely for hill breeds. A Herdwick grazing on the Cumbrian fells needs every bit of wool cover available to it, and a thickly fleeced tail naturally provides protection against the wind and rain whipping round its backside. In addition, the grazing available is relatively poor and there is little risk of the animal scouring, when the runny poo would soil the tail, attracting flies and therefore maggots, which would cause life-threatening damage – and there aren't many flies on mountain tops to take advantage in any case.

At the opposite end of the country, downland breeds grazing in lush southern pastures have a rich diet, and there are significant benefits to having a shortened tail when a combination of runny poo and opportunistic flies can cause a serious welfare issue.

Docking is carried out on sheep using the same rubber ring and elastrator as for castrating, to constrict the flow of blood to the tail. This must be carried out within seven days of birth (normally at the same time as castration), and sufficient tail must be retained to cover the vulva of a ewe lamb or the anus of a ram lamb. Apart from being illegal, avoid giving your lambs a tail as short as a bulldog's, because too short a tail can be a contributory factor in mature ewes prolapsing at lambing time.

DISPOSAL OF FALLEN STOCK

Fallen (dead) stock cannot be buried or burned because of the risk of disease spread through groundwater or air pollution. Instead, animals must be taken to, or collected by an approved knacker, hunt kennel, incinerator or renderer, either by private arrangement or through the National Fallen Stock Scheme. There is a charge for removing deadstock, but don't let this compromise your legal obligations for proper disposal.

Place your deadstock as close as possible to the entry of your property (contained and covered), so that the removal vehicle doesn't have to drive through your holding to seek out and collect the dead, as it is potentially something of a biohazard.

VEHICLES FOR THE SMALLHOLDER

Towing Trailers

Do check whether your car is suitable for towing a livestock trailer, and take account of the additional weight when loaded. A full car licence allows the towing of trailers weighing no more than 750kg, and you can tow heavier trailers with a car as long as the total weight of the vehicle and trailer isn't more than 3,500kg.

Quad Bikes and All-Terrain Vehicles (ATVs)

If you are using a quad bike (a sit-astride ATV) for agriculture, you need to register it with the DVLA as a light agricultural vehicle. An agricultural quad bike used on the road does not need an MOT or vehicle tax, but must be registered and licensed for road use, have a number plate and third-party insurance, and needs lights if used on the road after dark. Quad bikes used as light agricultural vehicles are not allowed to carry passengers – the long seat is to allow the operator to shift their bodyweight backwards and forwards for different slope conditions, and is not designed for passengers. If used for working, a suitable helmet is required. Riders should get proper training in the use of any ATV.

A side-by-side ATV (also known as a UTV, or utility terrain vehicle) is designed to carry two or four people and has seat restraints and roll-over protective cages. The same training as for quads is advised.

FURTHER LEGAL REQUIREMENTS

Farm Health and Safety Guidance

If you are running your holding as a business, your activities come within health and safety legislation and you must acquaint yourself with the requirements. You have liability for the actions of contractors and self-employed staff working on your holding. The Health and Safety Executive runs courses for farmers and publishes free guides to health and safety in agriculture.

Waste Exemptions

You may need to apply to the Environment Agency for certain waste exemptions such as chipping wood, spreading mulch, and many more possibilities.

Feeding Restrictions

It may be tempting to take your lunchtime leftovers to your goats or hens, but the general rule in the UK is that once any food has been in a kitchen (domestic, professional or industrial), it is illegal for it to be fed to livestock – and this extends to pets and those not intended for meat. This doesn't just refer to foods of animal origin, it also includes the vegetable portion, due to the risk of cross contamination, where products of animal origin such as milk are used in food preparation. Only if you live in a fully vegan household may you feed your kitchen scraps to pet livestock.

If you have a vegetable garden there may be veggie treats that your animals will enjoy, plus any suitable weeds you've patiently hoed. All you need do is take these straight from the garden to your pigs, goats, ducks and chickens without going through the kitchen first.

Planning Permission for Animal Housing

Planning permission for small-scale animal housing is not usually necessary, and permitted development rights exist for erecting structures in certain situations. Simple structures, such as temporary, mobile or other structures used for the purposes of agriculture, won't require permission. However, regulations will differ between commercial and hobby keepers related to the scale of the operation, so check with your local planning authority before you start any building work.

Animal By-Products Licence

If you wish to bring home horns or skins (to salt prior to tanning, or to take on to a tannery) from the abattoir from the livestock you have taken to slaughter, you need to register with APHA for the handling or use of animal by-products (ABPs); the form is available online.

Home Slaughter, the Abattoir and Selling your Produce

For information on the legal requirements for home slaughter, the abattoir and selling your meat, *see* Chapter 14. For the legal requirements for selling milk and dairy products, cider, jams and chutneys or eggs, *see* Chapter 16.

Cattle conservation grazing in culm woodland.

5 Tools and Equipment

It often concerns me that participants on our smallholding courses see all the equipment we have, and think that smallholding requires a stupendous budget for tools and kit, and massive barns to store it in. We do have a lot of equipment, but it has taken thirty-five years to accumulate, and most of the big items – a tractor and its implements, a 6-tonne digger, a skid steer and various trailers – have been through several owners before us. We started off with a fencing bar and a manual post bumper, both of which we still have, and a wheelbarrow that has long since rusted into oblivion and been replaced many times.

Decades on, I know that all these items were the right ones to get us started: we use wheelbarrows every day, and sorting out fencing is one of the very first smallholder priorities.

Hurdles can keep livestock in or out.

ESSENTIAL EVERYDAY ITEMS

What are the things that you are likely to need on a daily basis and from day one? Buckets are high on the list – for animal feed, for water, for carrying all sorts of things around (hammers, nails, screws, fencing staples, ducklings). And a nicely sharpened penknife with a handy length of baler twine should be permanently in your pocket. Then there are hurdles for containing and managing livestock (sheep, goats, alpacas) – and learn from our mistake: hurdle designs change over time and from one manufacturer to another, so buy more than you think you need or you'll end up like us, with four different designs, using baler twine to tie mismatched hurdles together.

If you are mucking anything out (poultry, goats, cows) you'll need a muck fork and shovel as well as that wheelbarrow, and spades are used not just for the vegetable garden but to dig out earth channels to sort minor floods and for planting those orchard trees. The range of gardening tools needed will depend on personal preference, but garden forks, a rake and a hoe get used a lot by us, as do loppers and secateurs, not just for trimming roses but for keeping gateways accessible, cutting back overenthusiastic brambles and willow, cutting browse for goats, and ivy to feed to any poorly sheep off its feed.

Cordless tools come into their own on the smallholding – who has an extension lead 500 metres long? A powerful (18-volt minimum) cordless drill with spare batteries is essential unless you want to spend hours with hand tools. Torches are another critical piece of kit: head torches free up hands to assist a lambing ewe in a dark corner of the shed, and being able to see what you're doing is helpful.

Don't forget various lengths of hosepipe. You may be very well organised and have taps exactly where you need them, but that is highly unlikely to be the case in the beginning, and you'll need to fill water buckets, water the veg, fill a wallow for pigs, a children's paddling pool for ducks and geese, wash mud and worse off your boots, every single day. Talking of boots, good wellies with a non-slip sole – and for preference neoprene lined to keep you warm in winter – plus waterproof jacket and trousers, many different types of gloves, and hats will be your new best friends. Forget about posh expensive waterproofs from hiking shops: go to your agricultural merchant and buy exactly what you need at a fraction of the cost (£25 rather than £250). And a boilersuit is the perfect option for keeping you warm and clean and providing plenty of pockets.

You'll need somewhere rodent proof to store animal feed: this should be metal rather than plastic, such as galvanised dustbins and old chest freezers (with the electric cord removed and any lock permanently disabled). And if you find the wildlife knocks open your dustbin lids to help themselves to the feed inside, use elasticated bungees to hold the lids in place. You'll also need feed scoops to dole out the rations, and you can buy these purpose made, or use empty fencing staple buckets or old ice-cream cartons.

FENCING ESSENTIALS

You can get contractors to put up new fencing but I don't know any smallholder who doesn't have to at least patch and mend fences from time to time. That iron bar, a post bumper, fencing pliers, hammers (sledge and hand), graft, and fencing strainers will all come in useful, plus the consumables such as staples, stock fencing and barbed wire. And then inevitably there will be a need for electric fencing of some kind – for keeping rabbits off the veg patch and the chickens inside their run, and the foxes out. Talking of pests, you will need an effective method of rat control, probably involving both cats and traps (*see* Chapter 18).

LIVESTOCK EQUIPMENT

I've already mentioned hurdles and buckets, but there are other items of equipment needed to keep animals in good shape: foot-trimming shears for sheep and goats, antiseptic spray, marker sprays, hand shears (to clip off dirty wool), ear tags and their applicators, appropriate housing – from hen huts to pig arks – strong detergent and disposable gloves. Once you are confident that you know how and when to apply medication of various kinds you'll need a stocked livestock medicine cupboard, complete with syringes and needles, spray and drenching guns, plus additional items in preparation for lambing, kidding, farrowing or calving.

Fencing tools.

Hand shears.

Piglets in a livestock trailer.

A little further down the line you might want to invest in an incubator to hatch chicks, keets, poults, ducklings or goslings (there are always good quality second-hand models for sale), and a livestock trailer to bring your animals home and later take them off to the abattoir. Depending on livestock numbers, you may decide in time that you want a sheep race, a weigh crate, or a turnover crate – but I wouldn't recommend investing in expensive, new electric shearing gear: sheep shearing is up there with coal mining for ultimately challenging physical labour. Use hand shears for trimming off mucky wool, but call in the professionals for annual shearing unless you are very fit and prepared to go on a shearing course.

This chapter covers the essential equipment, the things that will make your life easier if you have them, and those items of kit that are worth asking for on your birthday. Of course, what is essential to some will be entirely unnecessary to others. If you intend to raise birds artificially from hatching eggs, you can't do it in an airing cupboard, but if that's not part of your plan, you can dispense with budgeting for an incubator – and if you have no intention of dairying you won't need a cream separator or cheese moulds, or a honey extractor if you aren't keeping bees.

Water bowser to take water to livestock where there is no permanent supply.

BUYING SECOND-HAND

Many items can be bought second- or third-hand, but do your homework first on prices for new kit as some used things may be a false economy, costing almost as much as new and having a much shorter useful life. Items such as galvanised metal drinkers corrode and leak with age, which makes them entirely unfit for purpose, no matter how cheaply you bought them at the local car boot sale. All pre-owned housing, such as pig arks and poultry huts, should be given a very thorough disinfecting before allowing your livestock anywhere near it, as should equipment such as weigh crates, a sheep race and hurdles.

Everyday Essentials

Essentials for transporting and handling your livestock	
Crate	Kids, weaners or lambs will fit happily in a large dog crate in the back of a car. For poultry, cardboard boxes, cat carriers or dog crates work well.
Livestock trailer	For transporting larger livestock safely and appropriately you'll need access to a livestock trailer. Borrow, hire or buy this.
Hurdles	For containing goats, sheep, alpacas for handling, treatments, to create kidding/lambing pens. These come in 122cm (4ft) and 183cm (6ft) lengths, commonly 97cm (38in) high, although you can get them 1.3m (4.2ft) high for alpacas.
Halters, headcollars, collars and leads	For securing, moving or showing individual goats, sheep, cattle.
Housing, feeding and drinking	
Housing	From a little duck hut to a cow shed.
Feeders	Types of feeder depend on the species and type of food being offered, from chick crumbs to sow nuts and hay: small bale hayracks, large round-bale feeders, galvanised or plastic feeders and feed troughs, treadle feeders, rubber trugs.
Drinkers	Depending on species: shallow water buckets, nipple drinkers, galvanised and plastic drinkers, automatic bowl drinkers or ones with push tubes, cattle and sheep water troughs.
Buckets	Plastic, rubber, stainless steel depending on their purpose, for carrying anything and everything: milk, water, feed, tools, medicines. Calf buckets make ideal food and water buckets at lambing/kidding time.
Rodent-proof feed bins	Lidded metal dustbins or old chest freezers with the locks removed are cheap, effective options.
Feed scoops	Purpose-bought, or any container that you know holds the right amount of feed will do (such as large yogurt tubs, empty fencing staple buckets, ice-cream tubs).
Hosepipes	To get the water where you need it.
Mucking out and bedding	
Wheelbarrow	Plastic or metal.
Muck fork	For mucking out.
Shovel	As above.
Bedding material	Straw or shavings, or shredded paper.
Tools	
Spade	For all sorts of digging, particularly a narrow one with a metal shaft.
Hammer	Hand, club and sledge, for fencing, making and mending housing, and all sorts of repairs and DIY.
Crowbar	Lots of uses including repurposing pallets.
Saws	Pruning saw for keeping undergrowth in trim, and carpentry saws for woodwork.
Penknife	A penknife in every jacket pocket saves time and tempers (for opening feed sacks, cutting bale twine, cutting plastic).
Slash hook, loppers and secateurs	For keeping areas clear of scrub, nettles and so on, but only where necessary (don't be too tidy). To cut browse for goats.
Fencing tools and consumables	Iron bar, post bumper, wire fencing tools (such as fencing pliers, strainers), graft, staples. Even if you pay someone to do your fencing, there will inevitably be repairs, but hopefully not for many years.
Cordless drill	Not just drills but cordless anything for working in places where running electric extension leads is difficult or impossible. An extra battery is a good idea so you won't have to wait hours for a recharge while you're in the middle of a job.
Torch	Head torches in particular are brilliant on the hopefully rare occasions when you need to attend your birds at night, at lambing/kidding/farrowing or peering into dark corners.
Bolt croppers	For cutting quickly through wire if an animal is trapped.
Legal essentials	
Veterinary medicine record book	A legal requirement.

Holding register	A legal requirement.
Ear tags and applicator	For lambs/kids/calves born on your holding or to replace missing tags on adults.
Animal care and first aid essentials	
Dedicated storage cupboard for medicines and associated kit	So that you can find what you need immediately.
Digital thermometer	For taking temperatures rectally.
Foot shears	For foot trimming (only when necessary).
Dagging shears	For manual clipping to remove dirty fleece.
Faecal worm count packs (FEC packs)	To send away poo samples to check worm burdens, coccidiosis, fluke and so on.
Dosing and spray gun	To administer treatments.
Antiseptic spray	Use like a liquid plaster on minor wounds and foot problems.
Antibiotic spray	For foot infections and wounds where you are concerned about secondary infection (vet prescribed).
Syringes and needles	For administering injections.
Strong farm detergent	Many makes available. Use to clean sheds, bedding areas and equipment and as foot dips for biosecurity.
Sharps and doop (destruction of old pharmaceuticals) containers	For safe disposal of used needles and unused/out of date medication.
Stock marker spray	To mark animals that have been treated to avoid double or missed dosing, to match lambs and dams and highlight any problem stock.
Petroleum jelly and nappy cream	Soothing on legs, feet, sunburn and minor wounds (for livestock and you, but identify those for livestock use only).
Other essentials	
Somewhere dry to store straw/ bedding/hay	You don't want soggy forage or bedding.
A good vet	From a farm animal practice.
Freezer	For storing your home-reared meat, fruit and veg.
Other useful pieces of kit	
Reference books	*See* booklist in the appendices.
Egg trays	For storing eggs awaiting incubation or eating. Plastic trays last a lifetime, recycled paper ones can be composted.
Electric fencing kit	Batteries, energiser, tester, warning sign. *See* Chapter 15.
Rigid pond liner (for waterfowl)	Something that's not too heavy to tip out when full of dirty water. A redundant children's (water-tight) pool or sandpit makes a good alternative.
Incubator, candler, hatcher, brooder	For when you get into incubating eggs. Many types available – *see* Chapter 8.
Heat lamps and bulbs	For poorly lambs/kids, for growing chicks, ducklings, goslings, piglets.
Pest and predator prevention	*See* Chapter 18.
Chainsaw and associated safety gear	Helmet with visor, boots, trousers and gloves. To manage your own firewood a chainsaw will save hours and muscles, but please get all the necessary training and safety gear; this is not a piece of kit for a casual approach. If you are not competent, hire a professional.
Weigh crate	For managing animal health, medicine quantities and assessing meat return.
Footbath	For treating sheep and goats' feet.
Dairying kit	For goats, a milking stand, milk pail, strainer and filters, teat dip and strip cups, churn/milk container, dairy wipes and cleaners, disinfectants/sterilizer solutions.
Cheesemaking kit	Cream separator, cheesecloths, starter, rennet, cultures, stainless-steel pan, bucket, colander, thermometer, timer, moulds, trays, draining mat, press.
Beekeeping kit	Hive, protective clothing, smoker, hive tool, honey extractor, filters, jars.
Cider and juicing kit	Mill and press.
Polytunnel or greenhouse	To extend your veg growing season.

TRACTORS AND THEIR ASSOCIATED KIT

There will be people saying 'but what about a tractor?' or 'can't we at least justify having a quad bike?' – and who am I to deprive you of these most useful of vehicles? We certainly couldn't manage without a tractor and its various implements, in particular flatbed trailers (for carting hay bales), a tipping trailer (for collecting and depositing muck, earth and stone), a stockbox for moving pigs and sheep, all manner of haymaking kit, bale spikes, pallet forks and so on. But we really didn't need any of this for the first decade of our smallholding life: with 3 to 6 acres to care for we did everything by hand and wheelbarrow. Now, however, with 110 acres, many kilometres of Devon bank to maintain, and rather more animals to manage, we have had to increase our equipment bank. And an ancient Land Rover makes a very convenient mobile workshop out in the field.

Tractors, whether a compact type or something more farm-like, are often on the shopping list for new smallholders. Sometimes that's because the new place will be so much easier to manage with useful mechanical support, and just as often it's simply part of the dream and very much desired. When you're starting out, do you need to rush out and get a tractor, or do you resign yourself to failing miserably in your smallholding efforts if you don't have one? The answer is an unequivocal 'no'! The size of your holding, the tasks you want or need to do, and the livestock you have will all be considerations, as may be your level of fitness, your age and your budget.

4WD tractor.

Our first tractor purchase was the result of Andrew and a friend sitting in the pub one Saturday lunchtime and seeing an advert in a local farming paper. Next thing I knew, a Massey 135 – rather shabby but in perfect working order – was being driven off a flatbed trailer into our field. A tractor hadn't even been on our horizons, but sometimes it's best to grasp the opportunities that come your way. Twenty-five years later we still have that tractor, and it is used in good weather to pull small trailers, top thistles, rushes and docks, and to turn hay, move hay racks, poorly sheep and more.

Compact tractor.

If you are in wet parts of the country (that'll be most of the UK then), you'll be better off with a tractor that has a cab, as sitting on a sodden seat with the wind whistling round your ears and rain sliding down your neck might be acceptable for the occasional weekend, but is not advisable for daily use. Having said that, great care needs to be taken when the ground is soft so as not to compact it or

Massey 135 with hay bob.

create surface run-off – basically, there will be times of the year when you should not be using a tractor on your pasture at all (*see* Chapter 15).

Taking on cows was something of a game changer, and we really did need a four-wheel-drive tractor with a degree of heft about it – so we bought a twenty-year-old John Deere 6200 that could lift the larger hay, haylage and straw bales that we now needed to shift about on a daily basis throughout the winter.

If you want to make your own hay, or at least do some part of the process yourself, you'll need a mower, a hay turner/rake, and possibly a baler for your tractor. A local contractor may be happy to cut your grass and bale the hay at the end of the process, but if you are able to turn the grass with a turner (aka hay bob) once or twice a day to dry it, life is just that much easier for both parties.

Do consider a small farm tractor rather than a compact tractor; a small proper tractor such as our Massey 135, one of the David Brown 990 series, or a Ford 4610 (and there are many other options) will be cheaper than a compact (I'm talking second-hand here, as I know very few smallholders who go for brand new), will hold its value better, and the availability and subsequent cost of pre-loved implements will be far greater and easier on the pocket. Make sure the tractor has a standard three-point linkage, category 1 or 2, to which to attach the implements you need, and a drawbar for hitching up a trailer.

Trailers of all kinds – from flatbed to tipping to stockboxes – can be towed over rough ground by a tractor, and if your car isn't up to taking a livestock trailer down the road, the tractor will be – many farmers and smallholders turn up at the abattoir with their trailer on the back of a tractor.

Two-wheel tractor.

A front loader on the tractor means that you can pick up and move anything heavy, from dumpy bags of stone or cement, to a recumbent cow. You'll be able to push muck or soil about too. With bale spikes on the front you can move the larger bales of hay and straw, and if you find that you want to move things on pallets frequently, pallet forks for the front loader are a must.

Other tractor-powered equipment that smallholders might find helpful include a log splitter or wood processor (many smallholders rely on a log supply to keep warm in winter); a muck spreader if you have a few cows; a potato lifter (if you grow spuds on a fairly grand scale); a single-furrow plough to turn your ground; and a rotovator to prepare the ground for your veg.

All the pieces of equipment I've mentioned are readily available second-hand, and, just like the tractor, the value rarely drops if you buy well and keep them in good order.

Alternatives to a Tractor

Lighter and smaller types of mechanisation may be more appropriate than a tractor, particularly if you have wet ground. An ATV (all-terrain vehicle aka

UTV.

MAINTAINING FITNESS

The knowing smallholder becomes increasingly aware of the need to stay in a reasonable state of fitness. A bad back, for example, can put you in a very difficult position with regard to many smallholder chores, from digging up the spuds to checking your sheep's feet. Adopting no-dig gardening (good for the soil and the gardener) and using a hoe rather than bending down may be your approach, and there are tools and equipment that can save your body from getting into a twist in the first place. For five years we used muck forks and large wheelbarrows to muck out the cattle shed in the winter, but the resulting seasonal twinges led to a decision to buy a second-hand skid steer, and this has become a key part of our equipment. I'm also planning on finding a used turnover crate for handling the sheep before I reach my dotage.

quad bike) or UTV (utility terrain vehicle) – a sort of upgraded golf buggy that will tow a trailer, harrows, ploughs, aerators and tillers, and take you over all kinds of rough ground – could be just the thing. For those happy to have their muscles put to some use, a two-wheel tractor (aka a walking or walk behind tractor) with the appropriate attachments will rotovate your soil, plough, run a mower, raker and baler, brush cutter, chipper and log splitter, and even pull a trailer. It also has the added benefit of being able to get to those hard-to-reach corners or slopes.

A quad bike might suit perfectly as a means to get about and do the numerous smallholding jobs, and there are many possible attachments for light farm duties, such as trailers, mowers, tillers, feed snackers, rollers, paddock cleaners, weed wipers and sprayers. Just remember that the British weather is damp, so be mindful of year-round needs. And if you want to be completely green, how about a horse and cart?

Skid steer.

6 Fruit and Veg and Foraging

The idealist fantasy of a kitchen garden, particularly when walled with heat-absorbing and reflecting soft red brick, is enough to turn the most soil-averse soul into a theoretical gardener. Imagined versions are protected from winds, bask in sun, are bordered by jewelled flowers, and have secret doors suggesting mystery and adventure. However, the reality of the smallholder's fruit and veg patch may be less dreamy and more pragmatic.

I'm sure you know (or can work out) where to head if you want advice on the details of no-dig gardening, permaculture, crop rotation, companion planting, and all aspects of gardening sustainably, or to depress yourself about tree diseases and potato blight. This chapter simply sets out an encouraging approach to growing more of your own fruit and vegetables to enjoy the benefit of glorious plants that taste amazing.

THE ORCHARD

If you want an orchard, try to plant fruit trees in the first winter on the holding so that you will benefit from their crops before reaching your dotage. From late October to the end of March is the time to plant bare-root fruit trees. Choose young maiden whips (trees in their first year) if budget is all, but for those wanting a quicker crop, go for two- or three-year-old trees, known as bush trees. If possible, order trees from a local tree nursery; they will have varieties that flourish in your location. Although some trees are self-pollinators, for effective pollination you need apple trees that flower at similar times, so do check on-line pollination charts or talk to the nursery to choose complementary varieties.

Think about cider apples and perry pears, culinary and dessert varieties of apples, pears, plums and damsons, and there's nothing wrong with choosing trees that will probably feed the birds (cherries) before you get a chance to harvest the fruit, or are simply pretty. I planted medlars, not because I like to use the fruit particularly but because I find the trees charming, and mulberries because I love the fruit even though it will probably be several generations before anyone living here will get to taste one: they are painfully slow growing.

Our orchard is ridge and furrow, with the trees planted on top of the ridges to facilitate drainage on heavy clay soil. Avoid planting fruit trees in windy, unprotected positions, frost pockets, shade or areas of soggy ground. Depending on space, aesthetics and personal preference you might go for tall standard trees or dwarf trees, and this will be determined by the rootstocks you choose and the vigour of a specific variety, which in turn will dictate their spacing.

Old texts recommending planting fruit trees in big pits filled with manure are no longer considered correct, as this can create a water-logged hole in which the tree roots will die. Instead, remove turf and weeds from the planting site (we remove a metre square of turf and lay it upside down circled round the planting area), dig a hole and plant the tree with the graft union kept 5cm above the soil level. Add blood and bonemeal into the hole and backfill with the soil, treading in firmly.

Medlar.

Hazel nuts.

Damsons.

Rather than driving in a tall stake that can weaken the tree trunk, bash in a short stake with a rubber collar a few inches above ground to prevent the tree roots from damage by being rocked in the wind. Check the collar ties every year to make sure they aren't too tight.

Prune in the first four years to create the shape of the tree; a goblet shape letting in air and light

Orchard Tree Planting and Cropping

Rootstock	Planting distance between trees	Mature height	Crops from maidens	Full cropping age	Annual apple yield
Large standard M25	6-9m	4.5m	3-5 years	8-9 years	70-160kg
MM111	5m	3.5-4.5m	4-5 years	8-9 years	70-160kg
MM106 half standard and M26 semi-dwarfing	2.5-3.5m	2.5-4.5m	3-4 years	7-8 years	30-50kg
Dwarfing M9	2.5-3m	2.5m	2-3 years	5-6 years	15-20kg
Very dwarfing /miniature M27	1.2-1.5m	1.8-2m	2 years	4-5 years	7-12kg

Orchard tree guards.

is advisable. Old trees that are unproductive and crowded can be restored by vigorous pruning over several winters to avoid giving them too much of a shock. Apple, pear, quince and medlar trees should be pruned in the winter when dormant, but not in frosty weather, and plum and cherry trees in early spring.

Orchard Tree Protection

If you have livestock in an orchard or in other areas where vulnerable trees grow you will need to build tree guards around each tree. Livestock will chew the bark off even mature trees and kill them, but sheep or geese can still graze the grass in the orchard if the trees are protected. The only sheep breed unlikely to chew trees are the Shropshire, and that's not a guarantee, so make sure there is adequate grazing for them and be vigilant for any signs of nibbling.

If the guard you use is to be impenetrable by a goose, sheep or goat's head, it should be 1.8m (6ft) high and can be set close to the trunk; if the guard is of a design that allows livestock to get their heads through, it needs to be at least half a metre (1.6ft) away from the trunk, and a minimum height of 1.5m (5ft), using three or four fencing stakes to make a triangular or square guard.

To safeguard against rabbits you can avoid using plastic spiral rabbit guards, and source biodegradable collar guards and mulch mats to protect trees and the environment.

THE VEGETABLE GARDEN

Take time to determine the best site for the vegetable garden. Yes, of course it wants to be close to the house so that you don't fall back on tins because you can't face a trek to pick some vegetables for supper. Flat or south-facing is preferable. The quality of the soil is also a consideration, but in less-than-perfect situations build raised beds. All that muck and compost you'll be creating will over time hugely improve the soil structure. For rough, overgrown areas you might usefully put to work a couple of pigs or goats contained by electric fencing to do the initial clearing of weeds and unwanted shrub.

Our Christmas work involves sitting down with seed catalogues and deciding what we'd like to eat the following year. We keep notes of varieties we like that have been successful, and replicate those where possible, consigning to the 'no' pile vegetables that weren't much enjoyed or that didn't thrive. Ensuring that we don't have to survive on swede in the winter is not difficult: leeks, squashes (butternut in particular), Brussels sprouts, beetroot, carrots, parsnips, celeriac, celery, kale, onions, potatoes, cabbages of all colours and Swiss chard are easy to

The vegetable patch.

grow and will keep through the cold months, while there's no such thing as too many peas and they freeze beautifully.

So much summer veg can be frozen for winter: we grow all the makings for ratatouille (green and yellow courgettes, onions, red pepper and tomatoes), and bag and freeze large amounts, adding homegrown borlotti beans to the mix with plenty of fresh chilli for a vegetable chilli that can be enriched with our own minced beef or not, depending on the mood.

Your list of desired veg will be different from mine, but there are some things that are undeniably better when cooked or eaten moments after picking. Sweetcorn, broad beans, French and runner beans, asparagus, lettuces, cucumbers, tomatoes and herbs of every kind come into that category. I can't bear to buy fine French beans flown from Africa or South America, and only eat them when they are in season in the garden – but the season can be extended by the use of a cold frame, polytunnel or greenhouse.

I'd rather grow the things I love to eat in huge quantities as I'm happy to eat beans – yellow, green and purple – every day in salads or as a hot veg while they are at their best. You might prefer a wider range of veg planted in small numbers so that you can ring the changes, in which case you'll have to plan your planting so that you don't have everything ready to pick in the same week, leaving nothing for later.

Cauliflowers come to perfection and then beyond quicker than you can wink, so keep an eye on them and don't grow too many at the same time, whereas aubergines, peppers and chillies hang prettily from their plants for a good while before they start to turn. Courgettes (try planting yellow ones for their beauty and thin skins) are hugely prolific and need cutting every couple of days once they start to fruit if you don't want them to turn into marrows, and asparagus grows incredibly fast too and wants cutting daily.

Try growing things that you think you don't like; for example, fresh beetroot grated into salads or roasted with olive oil and a splash of balsamic vinegar is a world away from the sour pickled beets of childhood. On the other hand, nothing would push me into growing kohl rabi, a vegetable that has no place on my plate. You'll have your own dislikes.

Some things you love to grow will go past their best quickly – pak choi, coriander and rocket come in that category, so pick what you can and when they go spindly or start to form seeds, add them to the compost heap or feed them to the hens with thanks and no regrets.

Areas that are not cultivated or are being rested are best covered to suppress weeds: old carpets, flattened cardboard boxes, used silage sheets, wool shearings or weed-suppressant fabric are perfect for this. You can plant through weed-suppressant fabric or silage sheets, which helps the soil retain moisture.

Globe artichoke.

Leeks.

Yellow courgettes.

Aubergine.

Asparagus.

Rainbow chard.

Using sheep fleece for mulch.

Figs.

Gooseberries.

Strawberries.

THE FRUIT GARDEN

Strawberries, raspberries, blueberries, redcurrants, blackcurrants, gooseberries and other soft fruits are not difficult to grow; the difficulty is keeping the birds and squirrels off your delicious fruits. A fruit cage is ideal, but we've found that simply covering bushes with fruit netting also works. Considering the price of punnets of any berry in the supermarket, you can grow them at a fraction of the cost for a fairly extended season, picking autumn-fruiting raspberries grown outside in early November, and having strawberries as early as April when grown in a polytunnel or greenhouse.

POLYTUNNELS AND GREENHOUSES

Our polytunnel is 5.5 by 12m (18 by 39ft), giving a total area of 66sq m (710sq ft) covered, producing enough for us, our friends, family and course

Polytunnel.

Melon.

Sweetcorn.

Purple chillies.

participants. It showed no major signs of wear for ten years (with one door damaged in the wind over winter and the polythene just starting to tear), but the plastic cover didn't need replacing for three more years. It is used to grow not only delicate heat-loving items (tomatoes, cucumbers, melons, aubergines, squashes, peppers and herbs), but also to ensure the success of other crops, which, in a decent summer, can be grown outdoors (beans and peas).

It has as central bed 1.5m (5ft) wide), two pathways wide enough for a barrow, and two side beds each only 1m (3.3ft) wide. This means we can reach the centre of each bed without standing on the soil, though this might be seen as wasteful of space. However, when plants are at their rambling, magnificent peak they completely cover the pathways, so there are no regrets about the layout. We also have an area at one end for staging on which to grow seedlings. A separate greenhouse or potting shed would be nice to have.

Most sellers quote a price for the basic shell only, that is the frame and cover. They then offer a list of options, some of which are entirely essential.

Polytunnels are prone to wind damage so should be sited in a sheltered position, ideally south facing and unshaded. Consider concreting in post anchors if necessary. Avoid naturally dry spots; a damp area at the bottom of a slope may be ideal, provided you put down some hardcore paths for winter access. And don't position the polytunnel right next to a hedge that you intend to trim with a flail.

Irrigation is a major issue: remember that all the water will have to be supplied artificially, and due to the heat, daily watering is a must

Polycrub.

Proofing the seedlings against mice damage.

Polytunnel Options

Polytunnel options	Comments
Door kit	Lengths of 2×4 cut to size; easy to make your own, but make sure you have enough plastic to cover.
Anti-hotspot tape	This goes on the frame to stop it melting the plastic. *Essential.*
Crop bars	These fit across the hoops horizontally so you can suspend things, but also add to the strength of the frame. *Highly recommended.*
Timber base rail	There are two ways to attach the cover at ground level: dig a trench and bury it, or fit a timber rail to which the cover is attached. The rail option tends to leave gaps at ground level but it is easier to pull the cover evenly taut.
Overhead watering system	This requires 2-bar pressure – basically it must be mains water or a pump from a butt.
Side vent kit	Perforated plastic mesh can be used for polytunnel sides to improve ventilation, especially useful in hot areas.
Roll-up side kit	Allows the side of the polytunnel to be rolled up from a single crank handle.

in summer. A drip hose or automated watering system is a good investment; drip hoses are much more efficient and flexible and deliver the water to the plant root. We created a water harvesting system using the roof of a nearby building, collecting the water in a 5,000ltr (1,100gal) tank; this delivers water to the drip hose at very low pressure round the clock.

If you are in a very windy area, consider a Polycrub made in the Shetlands to cope with local conditions; made of polycarbonate rather than polythene, it won't tear in high winds or be carried off across country if properly anchored.

Weeds will thrive over the winter in the warmer polytunnel climate unless suppressed, so just like outside, use weed-suppressant materials. Mice tend to seek out the heat of the polytunnel so we created wire mesh covers to keep them off seedlings, in particular young pea plants, for which they make a beeline.

Our polytunnel is semi-circular, but if we were starting from scratch we would definitely go for one with straight sides to give more growing room for the area covered. We would also consider one with net sides (with a roll-up side kit so you can roll up the external polythene) to improve ventilation during very hot spells so plants don't burn up.

The greenhouse does the same job as a polytunnel but is smaller, tends to last much longer, and can be extremely attractive (and correspondingly expensive). If you are prepared to dismantle and erect it yourself, there are frequently second-hand greenhouses up for sale.

FORAGING

Blackberries, hazelnuts, rosehips, crab apples, gorse flowers, sloes (I don't like sloes, but you might),

Wild garlic.

PLANNING PERMISSION FOR POLYTUNNELS

Some development of polytunnels is allowed under existing permitted development rights, however, the local planning authority is responsible for deciding whether any type of planning permission is required for a particular development. The guidance states that the local planning authority gives appropriate weight to the agricultural and economic need for the development, and that circumstances where polytunnels can play an important role include: to provide protection for plants or young livestock, to secure improved quality produce, and to extend the growing season to provide greater opportunity for home-grown produce.

Gorse.

Wild strawberries.

Puffball.

rowan and elderberries and bullaces are glorious additions to your diet, foraged from hedgerows. Wild garlic proliferates in woodland and on banks. Beech nuts can be milled to produce oil, you may be in an area with sweet chestnuts (save some to add to your Christmas stuffing), wild strawberries grow in hotspots on grassy banks, and wild redcurrants grow in our woodland – although the wildlife gets to eat those.

Manage your hedges appropriately to ensure you don't deprive yourself and the wildlife of these free crops (*see* Chapter 15). Young beech and hawthorn leaves can be used as salad leaves, and every salad is beautified by adorning it with edible flowers such as primroses, violas, wood sorrel or nasturtiums.

Keeping all sorts of corners and patches wild is very much to be encouraged, and these neglected areas might provide you with fresh young nettles and sorrel to be enjoyed in soups, wild garlic for flavouring, and dandelion leaves for salads.

A word about foraging for fungi: some are unmistakable and safe to eat – the yellow chanterelle, for example, which is delicious, and giant puffballs as big as a football. We pick field and horse mushrooms growing in the fields, but if you don't know for sure how to identify varieties with certainty, think about growing from mushroom kits and logs inoculated with spawn from edible varieties, or buy spawn plugs and create your own mushroom-growing logs.

COMPOST

Perhaps this paeon to compost should have been at the head of the chapter. It's one of the delights of smallholding that you have the room and perfect materials for creating compost; you should never have to buy potting compost again, or be dismayed that your recycling caddies are filling up with vegetable matter. The mucky straw or shavings from poultry huts, and muck and bedding from any pigs, goats, sheep or cattle housed, however temporarily, will contribute a mass of goodness that will provide a

A relaxed approach to composting.

Compost bins.

lush growing medium for veg and fruit beds, once it has rotted down.

All veg and fruit waste, bits of woolly daggings trimmed off the back end of sheep, egg shells, tea bags, grass cuttings and leaves can be put into compost bins, or more casually heaped up where they can be allowed to rot down and later used to enrich the vegetable patch. Ideally have three areas or bins side by side so that you can allow one area to compost while adding new material to another, while the third is already in an advanced state or ready to use. Turning the heap from time to time composts it faster, and the more layered and varied it is, the better. Covering it with old carpet helps keep the heat in, killing off any weed seed, although do avoid chucking in nettles and docks and other pernicious weeds.

Depending on its contents and management, a compost heap can yield its black gold within six months; however, we expect to give it a year or even more.

7 Starting With Livestock

BUYING YOUR FIRST LIVESTOCK

Bringing home your first livestock is a major milestone, and not something to be rushed into. It's all too easy to go wild at a livestock auction and in the excitement of the moment bring home a trailer-load of unsuitable beasts. Resist this temptation: go and have fun, learn a lot, but leave your wallet at home.

Your first task is to decide which livestock you want to keep, and for what purpose. Are you keen to make your own dairy produce, be self-sufficient in eggs, desperate to avoid any more tasteless supermarket chicken, or rear everything you need to create the most delicious Christmas dinner ever? Are your smallholder intentions commercial or purely hobby related? Do you want to focus on sheep with fleece that is perfect for crafting, or are you after something to grace the field outside your kitchen window?

Light Sussex hen.

If you've enthusiastically turned straight to this chapter, do read the earlier chapters first as there are crucial requirements to put in place before exchanging cash for real live animals. For information on water, feed and shelter requirements *see* the relevant livestock chapter. For fencing, *see* Chapter 15.

Not All Breeds are the Same

Decide what you want from your livestock – eggs, milk, meat, fibre or wool, conservation grazing, show stock, companionship, joy, beauty, or a mix of these – and choose accordingly.

If cheesemaking for the home is your smallholding driver you will have to decide whether you would go for dairy goats or cows or milk sheep. To be pragmatic, there aren't many milk sheep in the UK so they may be difficult to source, cows are big and may be somewhat daunting at first, so a couple of dairy goats might be a more enticing and practical prospect.

For daily eggs, I'd always choose duck over hen, but you might disagree; in either case go for prolific egg-laying breeds, not something such as our large fowl Orpington hens that might squeeze out a paltry sixty eggs a year. Of the native breeds a Khaki Campbell duck will lay over 300 eggs a year, and a Light Sussex hen around 250 eggs, depending on the strain.

Just as geese are not simply ducks on a grander scale, and have different requirements, the same is true of sheep and goats: they are not interchangeable, although there are some commonalities. Goats will need access to shelter at all times, they have different feed, fencing, water and health requirements, and behave very differently to sheep, including their love of exercising their vocal cords. Nor are all pigs the same: the ginger lovelies with marvellous snouts that are the Tamworth are lively and are a great choice once you've built up some pig experience, so perhaps go for a more docile breed first; and the Mangalitza isn't ideal for pork chops but will provide all you could wish for if you want charcuterie and lard.

Choose what Gladdens your Heart

Spending time observing your animals improves husbandry skills; knowing and beginning to understand normal healthy behaviour will help you identify problems if they occur. If you choose livestock that

Goat heaven.

gladdens your heart and appeals to your eye you will spend more time with them, even if this mostly means gazing lovingly over the gate just watching them going about their daily business. If one breed of sheep or pig doesn't make you smile, choose ones that will. Even so, be practical too – a very large breed may feel overwhelming to handle, or huge horns may make you fearful, and the more primitive breeds tend to be flighty and more challenging to contain.

Age and Stage of Life

You really can start off with animals at any stage of life. From bottle-fed lambs, kids and calves or day-old chicks or ducklings, to all species on the cusp of breeding age, all the way through to broken-mouthed ewes and ten-year-old geese. I would advise the inexperienced smallholder to start off with point-of-lay birds (those just entering their first laying season), and with youngish sheep, goats and cows that have been through a successful pregnancy and birth, and possibly have their offspring at foot, too. For pigs, buy weaners at eight to nine weeks, unless you are committed to breeding, in which case an in-pig gilt or young sow would be ideal.

Where and How to Buy

Once your homework is done and preparations made, it's time to go shopping. Livestock are herd or flock animals, so you will need to buy a minimum of two of each species to ensure their health and happiness. This does not translate into buying two of absolutely everything – you are not Noah. If you are nervous of making a poor choice, overwhelmed by the options or the excitement of actual purchase, ask for guidance from an experienced keeper. Your best choice of seller is a reputable breeder, all of whom are keen to maintain their reputation and will be available to offer advice in those early months.

There are plenty of ways to find and meet breeders. Have a few days out visiting agricultural shows, and take the opportunity to discover more detail about the various breeds, and talk to the people showing their stock. Take a note of their contact details for later conversations; most have banners with this information alongside their livestock pens.

Many breeders have websites giving information about their livestock, and sometimes they list what is available for sale. The breed societies have websites, and a secretary who can signpost you to a local breeder. Farming and smallholding publications and websites will have classified adverts, as do online 'livestock for sale' sites and forums, and agricultural merchants still have the ubiquitous signboards with postcards of 'for sale' advertisements.

Once you have a shortlist of options, visit as many breeders as you can: you will get a good feeling about the ones you would like to buy from.

Livestock auctions are great fun to attend, and useful for developing familiarity with various species,

Sheep (Mules) at an agricultural show.

Alpacas at auction.

Judging Herdwick tups at Wasdale Head.

understanding what makes good locomotion (are they lame, or walking soundly?), differences in conformation and other traits. However, do be wary of purchasing from an auction, because even if the animals you are interested in are prime beasts arriving in the best of health, if they are penned next to some less-than-healthy examples they might be the recipient of disease or parasites by the time they leave the site. You won't be able to phone the auctioneer later for guidance on caring for your purchases, or asking any of the other things that you want help on.

That being said, do consider buying from dispersal sales or breed society shows and sales. At dispersal auctions everything is for sale, including all the animals that the vendor would be keeping if ill health, down-sizing, retirement, divorce, emigration or other significant life moments hadn't intervened. At society shows and sales, competitors are after rosettes and peer approval, so good quality stock is up for grabs, at a price.

BUYER BEWARE

Don't buy any livestock that you feel sorry for, as this is a sure road to misery and expense. Starting out on your livestock-keeping journey you want healthy, sound animals that will give pleasure, and buying poor examples in a dubious state of health only encourages unscrupulous sellers. Don't buy 'sight unseen' either: avoid collecting new stock from a motorway service-station car park, or having them delivered without seeing them at the seller's site first.

Visit the seller and assess if what you see is something you aspire to: not the glory of their house or garden, but how they are keeping their livestock. Don't expect pristine surroundings (there will be poo, probably baler twine and signs of pragmatic make-do and mend, and in wet weather there will be mud), but do check out the livestock, and how they are handled and cared for. As well as getting to grips with how well the stock is managed, you may pick up some helpful ideas for your own set-up.

If you are moving on to a smallholding, you might find the vendor keen to include their livestock in the sale. If you have any doubts about the animals' age, health or general fitness, or even if they simply aren't what you want, make sure that they are sold on before you take possession of the property. Disinfect any existing livestock accommodation and equipment with a farming and livestock approved disinfectant before bringing your own stock on site.

Purchase Costs

You can get animals at almost any price, from ridiculously cheap to outrageously expensive. Unless you are setting up a premium pedigree breeding business or going for the truly exotic, you don't have to pay way over the odds, but be careful of buying bargains. There's always a reason why something is cheap – mostly it's about offloading something the vendor doesn't want. Seller ill health, or change of domestic circumstances may force a quick cut-price sale, but that's no reason for taking something home with you without a thorough check, so ask questions and have a good look at your potential purchases.

The price of livestock varies hugely depending on species, breed, desirability, pedigree status, excellence of conformation, championships won, rarity, beauty, age, health, milkiness, meatiness and change in seller circumstances. If you are interested in showing your livestock you need to be buying examples of show quality, so be prepared to pay accordingly. It will cost you a handful of tenners for a few hens and thousands for a herd of cows, but whatever the species, the purchase cost is only the beginning when taking on any livestock: you need to have some financial resources to keep animals well, although for those handy with tools, many items can be made at little cost.

You can research current prices by looking at the online livestock sites, but if you have admired a particular vendor's stock be prepared to pay a premium as they will be in demand. On the other side of the coin, your homework on prices is important to do before committing to one person's livestock; I know of a number of new keepers who have paid significantly more (up to five times the going rate) than they needed to for breeds that are available at a more competitive price from multiple good sources.

Certain breeds will command a very high price, such as the Valais Blacknose sheep, because they are used in the same way that peacocks were in times past, to add visual interest to a lawn, rather than as a source of meat. If that's the case, you need to budget many thousands of pounds per ewe, rather than the more humdrum price of a couple of hundred for a desirable commercial breed.

SELECTING HEALTHY LIVESTOCK

Do visit the seller to see their stock on site, and be sure that you like what you see. If possible, buy from a closed herd or flock where they breed rather than buy in replacement stock, raising a percentage of their best young females to take the place of older ones that are no longer productive. This is desirable because every time new animals enter a holding they bring with them a disease risk. If home-bred females are being kept, unrelated males will have to be brought in, so be realistic about the nature of the closed status.

Ideally, buy livestock from as few sources as possible – the more places you buy from, the higher the risk of bringing in disease with your newly acquired animals. If your visit to a breeder rings warning bells – itchy, scrawny, ill-kempt stock with filthy bums and runny eyes for example – do not be their saviour unless you're setting yourself up as a rescue centre and are comfortable with significant and ongoing vet bills. Bringing home animals with a severe internal or external parasitic infestation can have a lasting effect on any stock you buy later, so start as you mean to go on with clear-eyed, smooth-skinned, lively, clean stock that you are proud to call your own.

What to look for:

- Are the animals structurally sound with no signs of lameness?
- Is the mouth well shaped without an undershot or overshot jaw?
- Do they have good body conformation? Are they overfat, too thin, just right?
- Is the coat/fleece/feather in good condition? Are there any sores or itchy patches? Are there any visible signs of external parasites such as lice, mites or mange?
- Are they nice and clean around their bottoms with no loose dung or scouring?
- Are they coughing or breathing fast? Are there any swellings/cysts?
- Are they behaving as you'd expect for the species: lively, active, not hiding in corners? Do they exhibit good temperament? Challenging animals are not for the inexperienced (possibly not for anyone).
- Is the breeder forthcoming and informative about their animals? Are they happy to share treatment history and medication that's been used in the last six to twelve months?
- Do they have any particular health problems on their holding? Do they vaccinate against clostridial diseases and/or any specific issues? Does their stock have contact with other people's livestock (through showing, neighbouring livestock)? Do they share equipment with other smallholders/farmers?
- Do they belong to a health accreditation scheme?
- Has the stock you are interested in given birth before, and was that successful? Are they pregnant? If so, when are they expected to give birth?
- What age are they? Are they the right age range for your needs?
- Ask if you can see the parents of the stock you are interested in. Check why they are for sale.
- Although not related to health status, if you want registered pedigree stock, make sure this is in place before you buy. Only the breeder can register their animals, and this is not something you can do yourself later.

If you are buying a ram, bull, boar or buck and want to improve certain aspects in the offspring (such as meatiness, conformation, soundness of feet, quality of the wool), focus on those traits in the male. Make sure he has really great feet and moves well, and don't consider anything that is lame, or is low on its pasterns (you want the legs to be upright, not low or sloping over the equivalent of ankles). Check that his teeth are in good order, with no staining round the mouth, and the teeth will also enable you to check his stated age. Testicles should be firm, contain two testes, and if possible check the penis to make sure it has no sores. He should be in good body condition, as he will lose weight once he starts mating. You don't want any sign of mites or lice, so check for a good, clean coat with no sores or rubbed patches.

Ask the owner for his vaccination and treatment history and how he has been fed, as this helps you plan your own ongoing husbandry. And in the excitement of the moment, don't just pop your new chap in with your females: quarantine him to monitor any ailments that might arise. Don't forget that your male will need company when not running with the females: a castrated male companion is ideal.

QUARANTINE

It's impossible to overestimate the importance of quarantine to avoid bringing disease on to your holding. Quarantine purchased animals for a minimum of twenty-one days (twenty-eight days is better), and watch closely for any signs of disease during that time. If anything looks suspicious, investigate fully by involving your vet. Even if the vendor has wormed the stock recently, talk to your vet about using one of the newer generation worm products to avoid wormer-resistant parasites contaminating your holding and creating future problems. They may, as an alternative, suggest worming with two different products, and you may also be advised to use a product to deal with any external parasites such as lice.

Quarantine facilities should ideally be somewhere that can be disinfected afterwards, or a small paddock that can be rested for several months after the quarantine has been completed. Don't quarantine anything in isolation: at the very least ensure they can see other stock of the same species, while making sure they can't physically contact each other.

Ageing a Ruminant

Teeth tell a ruminant's age, so it pays to understand what's going on inside their mouth; you won't want to buy animals at the end of their lifespan if you are hoping to keep and/or breed from them for many years to come. If you want to try your hand at breeding with an experienced ewe, doe, cow or sow, older animals may do you well as long as they are in good health and body condition, and don't have a history of difficult births or barrenness.

At the front of the mouth, cattle, sheep and goats have teeth (incisors) only on the bottom jaw, which meet a gum pad on the upper jaw. They have molars top and bottom at the side of the mouth; check these are in place as they are essential for dealing with their fibrous diet. Avoid animals with staining round their mouth as they are clearly having problems chewing – though don't confuse this with molasses stains or other temporary discoloration from licking a mineral bucket. Cattle will have a passport so their date of birth will be in writing. And no matter how hard you look, you won't find teeth in a hen's mouth.

Ageing Cattle, Sheep and Goats

Age	Teeth development (in the lower jaw) for sheep, goats and cattle as they age	
At birth	One or two temporary incisors.	*Newborn.*
Six weeks	Eight milk incisors.	*Six weeks.*
One year old (in its second year)	Two central milk incisors are replaced with larger permanent teeth (the animal is known as a 'two-tooth').	*One year.*
Two years old (in its third year)	It has four large incisors (and is known as a 'four-tooth').	*Two years.*
Three years old (in its fourth year)	It has six large incisors (and is known as a 'six-tooth').	*Three years.*
Four years old (in its fifth year)	It has eight large incisors (and is known as 'full-mouthed', or a 'full mouth').	*Four years.*
Aged	As the animal ages, the teeth will wear and it may lose some or all of its teeth, when it is called 'broken mouthed'; this can make the intake of adequate nutrition more challenging.	*Aged.*

Jaw Types

Jaw type	
Normal jaw. The incisors meet the dental pad exactly.	*Normal jaw.*
Undershot jaw. The teeth close behind the dental pad because the lower jaw is too short.	*Undershot jaw.*
Overshot jaw. The incisors extend forwards past the dental pad as the lower jaw is longer than the upper jaw.	*Overshot jaw.*

Check that goats, pigs, sheep and cattle have a normally shaped jaw, neither undershot (parrot-mouthed) nor overshot (monkey-mouthed), as either can make chewing difficult. In both conditions the incisors on the lower jaw fail to meet the dental pad on the upper jaw, meaning the animal is unable to get all the nutrition it needs. Those with this defect should not be bred from as it is a genetic trait that may be passed on to the offspring. In most species, the bite is set by the time the newborn is a few months old, and is unlikely to improve – the problem is likely to become more pronounced as the animal grows.

Udder and Teats

The udder and teats are of paramount importance if you are intending to breed from or milk a female. When checking an udder in anticipation of a purchase, feel for lumps that might indicate a history of mastitis (infection and inflammation of the udder), and therefore a subsequent inability to produce milk. Unless you are intending to show or breed for showing, an udder does not need to be a thing of beauty. However, it still needs to be functional, with well spaced teats that you can milk or that offspring can suckle from – so avoid enlarged, pendulous teats that a newborn can't fit in its mouth, or such tiny buttons that you can't hand milk. Furthermore, if you are planning to graze your livestock on rough land they don't want to be dragging a low-slung udder over thistles and scrub that would damage it, or be treading on their own teats; teats should be above the level of the hocks.

PEDIGREE STOCK

Pedigree stock are animals whose parentage within a specific breed is traceable, and has been registered with the relevant breed society (birds are not registered with breed societies). Only breeders can register their stock, and they must be paid-up members of the relevant breed society, or in the case of pigs and goats, the British Pig Association or the British Goat Society, as appropriate. Sellers sometimes refer to pedigree stock for sale as 'registered pure-bred', but 'pure-bred' alone simply means that offspring probably come from a sire and dam of a particular breed. If you want to play an active role in ensuring the survival of our rare and native breeds, buying an unregistered animal does not help that cause – a 'pure-bred' pig is just another pig, perfectly good for rearing and breeding, but it won't contribute to the continuation of rare breed bloodlines.

A pedigree is the recorded ancestry of an animal or bird, showing that its ancestors derive over many generations from a recognised breed – and the emphasis should be on 'recorded'. It doesn't matter if you know that your stock originated from pure-bred parentage going back generations: unless it has been registered as such in a herd or flock book managed by the relevant breed society, its provenance cannot be assured, and it is not considered pedigree.

SUITING YOUR LOCATION

Think about your location and its climate and terrain when choosing livestock. Keeping more delicate species, whether it's a lemon tree or an Anglo Nubian goat, on harsh terrain is not advised as they are unlikely to thrive. Feather-knickered poultry and those with feathered feet aren't going to like it much on boggy ground, and very hairy or thick-woolled livestock are at risk of attack by external parasites in humid climes. From the opposite perspective, livestock that come from colder, more dramatically weathered regions will have a relatively easy life in the softer south. Choosing breeds that are native to your area is always a good starting point.

EASE OF MANAGEMENT

Reams can be written on the topic of easy-care livestock. The truth is that not even the hardiest, least fussy of creatures can be brought home and ignored. Everything needs protection from the elements as appropriate to the species and breed – it also needs water, food, a healthy environment, the companionship of its own kind, fencing to protect it from hazards, and good husbandry to ensure that health and welfare is maintained and managed, and treatment given as necessary. It's true that some livestock are easier to keep than others and require less intervention, but your responsibility remains the same, with at least daily checks and visits.

This doesn't mean that you can't fit livestock into a busy life: as someone who worked full time, with a husband who spent significant periods working abroad from Monday to Friday, it absolutely can be done if you are well organised. Your fences need to be in tiptop condition, adequate amounts of feed need to be in stock, as does a first aid and medical kit, and good organisation is critical. It's no good receiving a call in the middle of the day or night saying your pigs have got out on the road when one of you is in a meeting 20 miles away and the other is on another continent entirely.

Having livestock with a challenging temperament and unruly behaviour is not going to help ease you in to the smallholding life. For example, buying cattle that have spent their life to date up on the moors and have been handled at most once a year is not a sensible choice for a first-time cattle keeper, no matter how docile the breed is meant to be. If you want hens that your children can spend time with, and collect eggs from without being pecked, there are plenty of charming docile breeds to choose from, so ignore those with a feisty reputation. Being clever about the livestock you choose, ensuring as well as you are able that they are fit for purpose and healthy, will give you the best experience.

TRANSPORTING YOUR NEW LIVESTOCK

I have seen more unsuitable methods of livestock transport than I care to remember, from the fully grown ram stuffed in the back of a hatchback, butting the rear window as it went on its way, to a plastic airtight lunchbox being offered as a carrier for chicks. If buying from a breeder, do ask them about the size of the animals you're buying, and get their guidance on appropriate transport and carrying containers.

Cat carriers make excellent bird transport containers: they have good ventilation, are easily hosed out, have a carry handle, and keep the birds secure. Large cardboard boxes work well for big birds such as adult geese, and it's easy to cut ventilation holes; just be sure to do this before you put the bird in. You can buy made-for-the-purpose cardboard animal carriers, but any sturdy cardboard box of the right size will do well for one-off journeys.

Large dog crates are suitable for a variety of livestock from birds to weaners, kids and lambs, as long as the spacing of the bars of the crate doesn't allow the animal or bird contained to escape, or to get parts of its anatomy trapped.

Transporting geese safely.

Livestock trailer.

For larger livestock you'll need a suitable roadworthy livestock trailer, which you can borrow, hire or buy. Some smallholders offer a trailer and driver service, which is something to hunt out when a few trips a year doesn't justify purchasing a trailer of your own.

Don't put feed or water in with the livestock, but do bring some for the journey just in case of delays. If you have animal containers in your car, keep them out of direct sunlight, and avoid transporting livestock when the weather is very hot or very cold. Take your new purchases straight home without any detours, and if you have to stop make sure it's brief and that the vehicle remains ventilated and is parked in the shade.

Sheep in a trailer.

LESS FAMILIAR LIVESTOCK

Unsurprisingly, considering how popular, useful and desirable they are, this book focuses on the core livestock that provide the smallholding with meat, eggs, milk, fibre and skins – but there are more options that deserve a mention. I haven't included game (venison, pheasants, wild rabbits, wild duck and the like), although wild foods are a welcome addition to the table.

Camelids

Alpacas and llamas may only rarely be used for meat in the UK – if you get a chance to try alpaca sausages

Alpacas.

they come highly recommended – but they have been popular for several decades now as guard animals, protecting poultry, sheep and goats from predators, including domestic dogs and foxes. Single castrated males or an unbred female integrated into your flock will act as a serious deterrent. Our llama (now long dead) would visit every newborn lamb in the field, sniff it and accept it into the flock for protection. He would chase loose dogs and foxes at speed, with long neck outstretched held close to the ground, and was an extremely effective guard animal. As well as guarding, alpacas have very desirable fleeces, of interest to the home crafter.

British national hives.

Bees

The joys of honey versus the fear of being stung is something that many will weigh in the balance, but beekeeping, whether in a rural or urban environment, is perennially popular. I am not a beekeeper, and would recommend contacting one of the 270 local branches of the British Bee Keepers Association; most offer beekeeping experiences and training courses, will provide a mentor, and give help with obtaining local bees.

Bees.

Beekeeping.

Rabbits

The concept of keeping rabbits for meat has rather wartime connotations of their being a poor man's meat, only fit for eating in desperate times, but anyone who loves to eat chicken but hasn't eaten rabbit should definitely try it. Delicious, lean and versatile, rabbits are, as everyone knows, productive: a doe can have from four to six litters in a year. Gestation is around thirty-one days, and she will wean her young at eight weeks.

In the same way as poultry and waterfowl, when a doe is ready to give birth she will pluck

Doe rabbit.

Rabbits in their hutch.

Kits in the nest.

her breast to provide a fluffy, warm bed for the newborn kits. Kits should not be handled by humans in their first week as there is a risk that the doe will reject them; once their eyes open at two weeks old they can be handled. At three weeks they spend time outside the nest, and by five weeks will eat pellets and drink water. Fed on a mix of commercial rabbit pellets and grass grazed through the base mesh of a movable run, they reach meat weight at twelve weeks, just before reaching sexual maturity.

Bucks should be kept separate from does once sexually mature, and you should only bring the doe to the buck for mating (not vice versa as the doe is very territorial and can severely damage the buck, even though he is larger). Does have no predetermined heat cycle, but will ovulate when put with the male. If you bring the doe back to the buck two to three weeks after mating and he ignores her, the chances are that she is pregnant.

Fish

Rearing trout in a pond or large water tank to eat fresh or smoked may not be a common smallholder pursuit, but it certainly adds diversity to your diet. Expensive to buy, this could be a real money saver for regular fish eaters, although you'll need to protect the fish from any passing herons. An air pump is needed to regulate the temperature of the water and keep it well oxygenated, and barley extract or barley straw to keep algae at bay. The fish will eat midges and insects that sit on the surface of the water, and will require additional feeding with trout pellets.

Rheas and Emus

Rheas and emus are great big birds, perhaps not in comparison with the ostrich (the latter requires a licence to keep as they are categorised as wild animals), but rheas stand as high as 1.5m (5ft) and weigh up to 40kg (90lb), and the emu may be 2m (7ft) tall and weigh up to 55kg (120lb); both birds have sharp claws. Although emus and rheas can be friendly and inquisitive, this is not always the case, and they are definitely not for everyone, or to be taken on lightly; nevertheless they are becoming increasingly popular in the UK as pets. You'll never outrun the rhea's top speed of 40mph or the emu's at 50mph! For feed, ostrich pellets can be ordered from your agricultural merchant or bought online; the birds also enjoy fruit and vegetables.

Rheas lay around fifty eggs per season and the incubation period is 36–44 days, while emus lay around twenty beautiful dark green eggs with an incubation time of 46–56 days. Natural incubation and chick rearing is carried out by the male. The meat is red and lean and is said to taste like beef.

Water Buffalo

All the rules relating to keeping cattle (*see* chapters 4 and 12) are the same for buffalo. In addition to providing beef they have the added bonus as the perfect dairy beast for mozzarella makers. They are known to be tame, curious animals,

Rhea.

Emu eggs with a hen's egg.

quieter to milk than most cows; however, when they calve they are very protective and great care must be taken around them at this time. There are several large and small water buffalo herds now in the UK, kept for milk and meat; the meat is leaner than regular beef. More like pigs than cattle, water buffalo create wallows to coat themselves in mud to regulate their temperature, and will spend significant time in wet mud and water, so they may be of interest if you have natural ponds.

8 Poultry and Waterfowl

Poultry and waterfowl are very much the smallholder starting point, and frequently an urban interest in chicken keeping is the seed from which a desire to be a more ambitious smallholder grows. The delight of scooping a warm, freshly laid egg out of the nest never wanes, and the mind travels on from that satisfying trio of hens to perhaps a few ducks and geese, and the possibilities of hatching chicks.

You'll need to provide housing that is well ventilated but not draughty, fresh bedding as needed, with a perch for chickens of the right dimensions for your breed. It must be rat-proof, fox-proof, mink-proof and badger-proof, so rotten, flimsy materials are not acceptable, and be prepared to disinfect everything as needed to protect your birds from external parasites such as the dreaded red mite (*see* Chapter 13). You'll also need to provide an outside area for the birds to spend their days, and unless you spend every hour of daylight in the garden as a deterrent, this means the creation of a fox-proof run, or if you opt for free range, accepting the inevitable losses. Doling out fresh water and food every day will be part of your commitment, as will never failing to lock them in their huts before dusk.

Duck hut.

BIRDS FIT FOR PURPOSE

Egg-laying Birds

If you want to be self-sufficient in eggs, choose from egg-producing or dual-purpose chicken breeds – and don't forget the joy of duck eggs. For geese, there is less of a variation in egg-laying abilities, with most laying around thirty eggs a year, although the prolific Chinese goose can achieve up to eighty if you're lucky. You may think of turkeys and guinea fowl primarily as meat birds, but their eggs make perfectly good eating. As for quails, their beautiful little eggs are delicious and make for a spectacular offering at the dinner table.

Buying birds at 'point of lay' means that their first laying season is imminent. Chickens start to lay at sixteen to twenty weeks of age, depending on the time of year and breed; if they hatch in late summer they may not produce their first eggs until the following spring. Ducks reach point of lay between sixteen and twenty-eight weeks of age, while geese tend to start laying during the spring following their hatching.

Collect eggs daily and store them somewhere cool (a fridge is unnecessary), using the oldest first so that you don't rack up steadily ageing eggs. They stay edible for several weeks, and the method of putting them in water to test for freshness works well: an egg sitting flat in the bottom of a bowl of cold water is fresh; it will start to tilt upwards as it ages and the air sac expands; and if it floats it wants chucking away.

Quail eggs.

Meat Birds

For meat, choose dual-purpose or table chickens and duck breeds. For turkeys, my favourite bird is the Norfolk Black – the most delicious turkey meat – but choose from any of the rare breeds. If you haven't eaten guinea fowl you are in for a treat, and any of the colour varieties will provide a good meat bird (live weight is around 1.5–2kg). All domestic geese make excellent eating, but there is a significant difference in weight and stature between breeds.

If you like the idea of producing your own poultry meat, you'll have to learn how to dispatch your birds humanely, pluck them neatly, and dress them hygienically (*see* Chapter 14).

Birds as Pets

Are you less interested in the production side of the birds but want them more as pets? Hatching your own or getting very young birds will enable you to handle them and get them fully acclimatised to being happy around humans, and choose breeds known for docility. Chickens that fit that role include Brahma, Cochin, Croad Langshan, Dorking, Orpington and Silkie, but are not limited to these. Ducks

Aylesbury duckling.

that are particularly tame include Pekin, Aylesbury, Khaki Campbell, Call and Muscovy. Docility is often determined by gentle handling and time spent with them, rather than the breed, although there are flighty, nervous exceptions.

Hand-rearing goslings is essential if you are hoping for pets. However, be aware that in the breeding season there is no chance of peaceful goose docility. I advise against having geese if you have young children unless you can guarantee keeping them apart.

Rare Breeds

If playing your part in conservation is of interest, you have plenty of choice across the poultry and waterfowl spectrum; check out the annual watchlist produced by the Rare Breeds Survival Trust.

CHICKENS

Thirty-five million hen eggs are eaten in the UK every day, and twenty million chickens are slaughtered for meat each week. Chickens and their eggs are absolute staple diet items, but there is much concern about welfare, provenance and over-industrialised poultry farming. Smallholders are in the enviable position of being able to set up their own egg and poultry supply chain, with food miles measured in metres, absolute provenance and complete control over welfare.

There are more than a hundred breeds and hybrids in the UK, and many different coloured varieties for some breeds. Dual-purpose birds provide plenty of eggs and are substantial enough to produce a good

Popular Chicken Breeds

Breed	Description (egg numbers per annum, live weight)
Dual-purpose birds	
Australorp	Black, white or blue feathers. Lays 200 brown eggs, meaty body of medium size. Bantam and standard sizes. Cock 3-4kg (bantam 1kg), hen 2.25-3kg (bantam 0.75-1kg).
Barnevelder	Black, double laced and partridge colouring. Hardy, lays 180 brown eggs. Cock 3-3.5kg (bantam 1kg), hen 2.5-2.75kg (bantam 1kg).
Brahma	Large, docile, heavy meat bird with feathered legs in dark, light, white, gold, buff Columbian varieties; lays 140 brown eggs. Cock 5.5kg, hen 4.5kg.
Croad Langshan	Large, black docile bird; lays 140-200 pink-brown eggs. Feathered legs and feet. Cock 4kg, hen 3-3.5kg.
Derbyshire Redcap	Hardy, active, fine flavoursome table bird; lays 150-200 white eggs. Rose comb, orange brown with black tails. Cock 3.5kg, hen 2.75kg.
Dorking	Large docile bird, popular as pets, in dark, white, cuckoo, silver grey and red colourways; lays 150 white eggs and has a fifth toe. Cock 3.5-6kg (bantam 1.2kg), hen 3.5-4.5kg (bantam 1kg).
Faverolle	Hardy, placid with a muff, beard and five toes, in salmon, black, buff, ermine, blue, cuckoo and white, laying 160-200 brown eggs. Cock 4-5kg (bantam 1.2kg), hen 3.5-4kg (bantam 1kg).
Maran	Friendly active bird in black, copper-black, dark cuckoo, golden cuckoo and silver cuckoo; lays 150-200 very dark brown eggs. Cock 3.5-4kg, hen 2.5-3kg.
Orpington	Friendly, docile, big bird, makes good pet and an excellent meat bird in blue, buff, black, white, jubilee, spangled or cuckoo colour varieties. Lays 60-160 brown eggs. Cock 3.5-4.5kg, hen 2.7-3.6kg.
Rhode Island Red	Red to black medium/heavy breed, lays 180-300 brown eggs. Cock 4kg, hen 3kg.
Sussex	Popular bird with eight colour varieties (light, speckled, brown, silver, white, buff, red and coronation). Lays 180-260 brown eggs. Cock 4kg (bantam 1.5kg), hen 3.2kg (bantam 1kg).
Vorwerk	Hardy and active, buff body with black head, neck and tail, medium-sized; lays 170 cream eggs. Cock 2.5-3kg, hen 2-2.5kg.
Wyandotte	Pretty colour varieties including barred, black, blue, blue-laced, blue partridge, buff, buff-laced, Columbian, gold-laced, partridge, red, silver-laced, silver-pencilled and white. Medium-sized, lays 180 brown eggs. Cock 3.5-4kg, hen 2.5-3.2kg.
Egg layers	
Amber Lee	Friendly, white-cream hybrid producing 330 brown eggs.
Ancona	Nervous, flighty, active small bird, black with white spots, lays 160-180 white-cream eggs.
Andalusian	Blue, black or splash with black lacing, early layer producing 160 white eggs.
Araucana	Hardy and flies well, crested with ear muffs and beard in twelve colours in large and bantam sizes. Lays 180-250 blue or green eggs, and uniquely the colour runs throughout the shell.
Bluebell	Dark blue-grey feathering, Rhode Island Red × Maran hybrid laying 250 brown eggs.
Bovan Goldline	Calm, docile Rhode Island Red × Light Sussex hybrid; lays 300 brown eggs.
Cream Legbar	Autosexing, hens have cream necks, salmon chests and grey-barred bodies; lays 180-200 blue, green or olive-blue eggs.
Minorca	Active and hardy, in black, blue and white varieties, lays 170-200 large, chalk-white eggs. Hen's comb folds to one side and is susceptible to frostbite.
Welsummer	Popular bird with golden brown head and hackles and red-brown body; lays 180 dark brown eggs.
Table birds	
Bresse	The white meat bird with blue legs favoured by the French; lays 250 cream eggs. Not docile. Cock 3-3.6kg, hen 2.2-2.5kg.
Cornish (Indian Game)	Hardy, muscular, protective bird, dark, Jubilee and double-laced blue colours. Lays 80-100 light brown eggs. Cock 3.6kg (bantam 2kg), hen 2.7kg (bantam 1.5kg).

Breed	Description (egg numbers per annum, live weight)
Ixworth	White rare breed with high quality meat, lays 150–180 white-cream eggs. Cock 3.5–4kg, hen 2.7–3.2kg.
Ornamental birds	
Ayam Cemani	Completely black: feathering, comb, tongue and meat, lays cream eggs.
Cochin	Docile, popular as pets, very large, fluffy, long-lived bird, feathered feet, in black, blue, buff, partridge and white. Lays 120 brown eggs.
Frizzle	Hardy, good layer, with curled or frizzled plumage, producing 120 white or pale brown eggs.
Hamburg	Active fliers, excitable and not suitable to keep with young children. Colours include black, gold-pencilled, silver-pencilled, gold-spangled and silver-spangled. Lays up to 150 small white eggs.
Old English Game	Hardy, active, noisy, dislikes confinement, comes in many colours; lays 40–120 pale brown eggs.
Silkie	Small, fluffy bird in various colours with black skin and bones, blue earlobes, and five toes. Lays 80 cream eggs, often used as broody for hatching.

Black Orpingtons.

Chickens free ranging.

meaty carcase. There are heavy breed chickens where the female hen averages more than 2.5kg (5.5lb) at maturity, light breeds weigh less, and bantams are miniature versions of their larger breed counterparts, or small standalone breeds. Most egg breeds tend not to go broody, so you will need an incubator or a surrogate broody bird if you wish to hatch.

I strongly encourage smallholders wanting quality meat to disregard the mass-produced hybrid meat birds (Hubbard, Sasso and Ross/Cobb) that will produce meaty but fairly tasteless carcases of the type available in every supermarket, raised commercially for a mere thirty-two days. Instead, opt for one of the rare or native breeds, and rear them for at least twelve weeks (we take them to twenty weeks or more) for a far more delicious result; choosing this approach is probably at the heart of why so many smallholders pursue this way of life.

DUCKS AND GEESE

I've written a whole book about keeping ducks and geese, which probably tells you just how much I appreciate waterfowl. Duck eggs are my breakfast joy, and both birds produce the best celebratory meat feasts – but if you are used to plucking a chicken, ducks and geese can take a whole lot longer, with down as well as feather to contend with.

Geese are not simply large ducks, and their behaviour, personality, egg production, impact on the land and more are very different. As grass eaters, geese create a thick, neat sward, mucking copiously as they go, cropping the grass to an even height, creating grassland that looks more like a lawn than a paddock. Their bills are the equivalent of a toddlers' fingers – into everything and quite powerfully destructive if allowed access to anything you want to keep in one piece, so protect anything and everything that you want to keep out of a goose's reach.

Ducks have very different sward habits: they make holes in wet ground, and with determined bills continue working at it until there is no more grass and

Call ducks.

instead plenty of mud. If you can move them, the grass will recover well (it has, after all, been generously fertilised with duck poo), but ducks don't so much create a lawn as damage it. Both species appreciate a pond, but as long as they have fresh water daily to drink and deep enough to dunk their heads under, they will manage just fine.

If of similar size and preferably reared as a group, ducks and chickens can be housed together, but do put a droppings board (hardboard works well) underneath the hen perches to stop the chickens pooing on the ducks at night. Although young goslings can be kept together with ducklings and chicks, maturing geese should be kept separately from other species,

Popular Duck and Goose Breeds

Breed	Description (egg numbers per annum, live weight)
Ducks – dual purpose	
Abacot Ranger	Lively foraging duck. Easily sexed from eight weeks of age, the male has an olive-green bill, the female's is dark slate grey. Good broodies, producing 180–200 eggs. Long-lived and placid. Drake 2.3–2.5kg, duck 2–2.3kg.
Magpie	Welsh breed, with black and white, blue and white, and dun and white varieties with orange legs. Flies fairly low to the ground. Lays up to 200 white to pale blue-green eggs. Drake 2.5–3.2kg, duck 2–2.75kg.
Muscovy	Not true ducks (perching waterfowl); they forage grass and eat large numbers of insects. Females are half the size of the males. Many colour variations, with red facial patches and caruncles. Broody, and will rear their young. Fly well and roost in trees/on roofs. A quiet breed. Lay 100–180 pale olive to white eggs. Incubation 35 days rather than the usual 28 days for ducks. Meat is highly prized. Drake 4.5–6.3kg, duck 2.3–3.2kg.
Saxony	Large, active bird. Bill yellow, legs orange. Egg colour white. Lays 80–100 eggs. Not a broody breed. Noisy females. Drake 3.6kg, duck 3.2kg.
Silver Appleyard	Cross-bred from Rouen, Pekin and Aylesbury; lays 160 to 250 white eggs, and a quick-growing table bird with a deep meaty breast. Drake 3.6–4.1kg, duck 3.2–3.6kg.
Welsh Harlequin	Outstanding egg layer. Drake's head is greenish-black, duck has a creamy-white head with brown stippling. Good foragers, they produce lean white meat and lay 240–330 white eggs. Drake 2.3–2.5kg, duck 2–2.3kg.
Ducks – egg layers	
Black East Indian	Bantam with black beetle-green feathers. Bills, legs and webs are black. Excellent fliers so will need their wings clipping. They lay 40–100 grey eggs that fade to white as the season progresses. Drake 0.9kg, duck 0.7–0.8kg.
Campbell	Campbells rarely fly, and come in khaki, white and dark colours. The Khaki Campbell is the most successful egg-laying utility breed, producing an astonishing 300–350 white eggs. They are of no use as broodies. Good quality, lean meat. Drake 2.3–2.5kg, duck 2–2.3kg.
Indian Runner	Unique upright stance, long and slender. Fourteen colour variations in the UK. Popular as pets. Not a flying breed. Very moderate appetite. Lay 200–250 blue-green to white eggs (early eggs from black runners can be very dark grey). Drake 1.6–2.3kg, duck 1.4–2kg.
Shetland	Critically endangered Shetland Islands native. Tough, active birds. Black with green/blue sheen with white breast and occasional spotting. Drakes have yellowish bills and orange on their legs, the females have grey or black bills and legs. Lay 150 white to grey eggs. Drake 2kg, duck 1.8kg.
Ducks – table	
Aylesbury	White in colour; the exhibition Aylesbury has a pronounced keel and long pink bill with bright orange legs and webs. Lays 80–100 large white eggs (eggs are often blue in the USA). Drake 4.5–5.4kg, duck 4.1–5kg.
Pekin	Upright penguin stance (German Pekin – the American Pekin is nothing like as upright). Thick neck and white plumage with a yellow sheen, yellow bill and orange legs. Exhibition breed and important for commercial meat breeding. Lays around 200 white eggs. Drakes 4.1kg, ducks 3.6kg.
Rouen Clair	Traditional mallard colouring. Tendency to get fat, so need room for foraging and exercising, and avoid overfeeding. Docile, and do not fly well. Lay 150–200 pale blue-green to white eggs. Drake 3.4–4.1kg, duck 2.9–3.4kg.
Ducks – ornamental	
Call	The smallest domestic duck; highly vocal. Seventeen colour variations plus crested types. Excellent flier so will require clipping. Although small it has tasty meat. Lays 20–60 white eggs and makes a decent broody. Drake 0.6–0.7kg, duck 0.5–0.6kg.

Breed	Description (egg numbers per annum, live weight)
Geese – dual purpose	
Chinese	Smaller than the African with much larger knob. Long, slender neck. Brown, grey and white. Lays 40–80 white eggs split into spring and autumn seasons. Good guard geese and voluble; not safe around small dogs. Susceptible to cold (in particular the knob) so need shelter from very cold weather. Can fly. Gander 4.5–5.4kg, goose 3.6–4.5kg.
Embden	The tallest goose, standing over a metre. Pure white with light blue eyes, orange bill, legs and webs. Protective breed, not suitable around small dogs or young children. Lays around 30 white eggs. Gander 12.7–15.4kg, goose 10.9–12.7kg.
Pilgrim	Auto-sexing breed; the female is completely grey apart from her stern (backside) and spectacles, the male is mostly white with touches of pale grey on its back, wings and tail. At hatching the females have a dark bill and darker down than the males. Lay about 30 eggs, and make good broodies and parents. Gander 6.3–8.2kg, goose 5.4–7.3kg.
Roman	Pure white, small, chubby birds. There is a crested variety. Can be nervous, so interact with them when young to calm them. They are excellent layers, producing 40–65 white eggs. Gander 5.4–6.3kg, goose 4.5–5.4kg.
Steinbacher	Small, attractive goose in blue, grey, lavender, buff, cream and white varieties; the bill has a black bean at the tip and black serrations. Known as a fighting goose, but no more aggressive than other breeds. The female lays 5–25 eggs and has an unusually short breeding life – she is unlikely to lay beyond six years of age. Gander 6–7kg, goose 5–6kg.
Toulouse	Very large; grey and buff and white varieties. Exhibition birds have a deep keel and large dewlap, massive heads and short legs. Their size can make it challenging to breed. Utility Toulouse are slighter, with no dewlap and are invariably crossbred. Lay 30–40 white eggs, although some strains produce an astonishing 220–240 eggs per year. Susceptible to flystrike. Excellent for crossing to produce meat birds, as pure Toulouse are very fat-laden. Short lifespan, averaging ten years, although they can live for twice as long. Gander 11.8–13.6kg, goose 9.1–10.9kg.
Geese – egg layers	
Czech	The smallest domestic goose, lively nature, white with blue eyes, orange-red bill, legs and webs. The female is shorter than the male with a thicker neck and deeper body. Rather noisy; make good broodies and parents. Lay 40–60 eggs. Gander 5–5.5kg, goose 4–4.5kg.
Geese – table	
African	Large bird standing a metre tall, with prominent knob at the top of the bill. Surprisingly gentle for its size. Colour variations are brown, grey, buff and white. Meat is lean. Lays 20–30 white eggs, and goes broody. Gander is 10–12.7kg, goose 8.2–10.9kg.
Pomeranian	Plump, meaty body. Most common colour in the UK is grey and white pied/saddleback; there are also all-grey and all-white varieties. Hardy breed and good forager. Lay 30–40 eggs; make good guard birds. Gander 8.2–10.9kg, goose 7.3–9.1kg.
Shetland	Autosexing; the males are mainly white, the females grey and white. Small and hardy, a quiet breed. Good foragers used to rid the grass of parasites, such as liver fluke. Fly well and lay around 20 eggs per year, but do not make a reliable broody. Gander 6kg, goose 5kg.
West of England	Autosexing; the male is mainly white with some grey on the back and rump, the female has a saddleback grey and white pattern, with grey on the head and neck. Can be sexed at hatching: the female has grey patches on the beak, plain pale orange for the male. Lays 20–30 white eggs. Gander 7.3–9.1kg, goose 6.3–8.2kg.
Geese – ornamental	
Sebastopol	Flightless bird with uniquely frizzled/curled feathers, in two varieties: the curled feather and the smooth-breasted (or trailing feather). The majority are entirely white, but there are also grey-flecked, grey, saddleback and buff versions. They need access to swimming water to care for their unusual plumage. Not particularly docile. Prone to wing abnormalities. Produce 25–40 eggs. Gander 5.4–7.3kg, goose 4.5–6.3kg.

Pilgrim geese.

and definitely before they enter the breeding season, when they can easily kill smaller birds. Even adolescent goslings can clumsily step on and squash smaller birds, so separate them as soon as they become disproportionately larger than their hatching mates.

TURKEYS

At the top of the domestic poultry weight scale comes the turkey. Temperamentally different from chickens, they are a little more challenging to herd, so take a degree of patience when putting them to bed as they come towards you rather than moving away from you. We buy in young birds, and keep turkeys for the time required to take them from month-old poults to meat

Norfolk Black turkey stag.

size (from June to December). They have an excellent food intake-to-weight gain ratio, eating sparely and growing well.

They are amusing, curious creatures, but notwithstanding their size are equally prey to foxes and need to be kept safe, as do all poultry. Turkeys should be kept on ground that has not been used by other poultry. They do grow large, so a good-sized run is required. They fly surprisingly well for their bulk, hopping up and over five-bar gates with no problem, so take this into account.

GUINEA FOWL

Guinea fowl are similar in size to a medium chicken, and their meat has a wonderful flavour. These balloon-shaped, polkadot-feathered birds are renowned for their guard-dog qualities, and make a noisy cackling when someone approaches. They eat ticks, which is a useful trait, and are very flighty, loving nothing better than to spend their time in trees or on a roof.

Young guinea fowl keets will fly almost from day one, so if buying them to rear, make sure their enclosure is tall and secure, with netting over it. I once put some day-olds into a workshop to rear under heat, and minutes later they had all flown out over the top of their enclosure and were hopping about the tools, hiding in curls of wood shavings and under the workbench. The seller told me he'd put in a few extra, and it was some time before I was sure that all were safely contained in the hurriedly improved set-up.

By the time guinea fowl become highly vocal they are ready for the pot, so you might prefer to buy in young birds to rear, rather than keep breeding fowl if you have neighbours or dislike their racket. They are keen foragers and will find much of their own food if free ranging, but to gain weight will still need access to poultry feed. The only way to keep them in a run is to roof it with netting or similar.

Guinea fowl.

Guinea fowl keets.

QUAIL

Quails may be tiny birds, but their eggs and meat make up for their lack of size in quality, and they are a real delicacy. The eggs are beautifully speckled and delicious, and are of niche interest.

Although many breeds live on the ground, some breeds of quail can be flighty, so they need to be kept enclosed in a rodent-proof aviary or in large adapted rabbit hutches. They have long toes and like to perch, so provide perches. Female birds live together happily, but the males fight and need housing separately. Coturnix quail can bash their heads on the roof of their hut, so stretch fruit netting below the roof to stop them injuring themselves if they take off in fright.

Quails start to lay eggs as early as six to nine weeks, depending on the breed, producing one egg a day. Meat birds are normally slaughtered at forty to forty-five days; adult weight is reached at about fifty days. Commercial quail feed is available.

Quail.

THE POULTRY KEEPER'S YEAR

For the casual poultry keeper, who rears small numbers of birds for meat and eggs, the seasonal outline in the poultry keeper's calendar should give you a sense of what to expect.

The Poultry Keeper's Calendar

Month	Activity
January	In the winter months you might despair of ever seeing another egg depending on the species, breed and age of your birds. Ensure water drinkers aren't frozen. Repair any defects in huts and runs. Test and thoroughly clean incubators, heat lamps and brooders ready for hatching. In very cold weather frost bite on chicken combs is a possibility, which can be avoided by rubbing on petroleum jelly.
February	As January, and keep on top of rodent problems. Birds will be starting to lay; test for fertility if you intend to hatch or to sell hatching eggs.
March	Incubation (under a broody or in an incubator) starts in earnest. Young hatchlings from incubators will be kept under heat for up to four weeks. Greenery from the veg garden can be added to birds' diets.
April/May	As March, and young birds can be put out on grass in covered, protective runs. As the weather heats up make sure huts are well disinfected and birds treated if necessary against external parasites/mites.
June/July	Make sure birds have shade as temperatures increase, and keep drinkers clean and refreshed. Continue to be vigilant for internal and external parasites and rodents.
August/September	Egg laying for some breeds/species will decrease significantly. Some birds will come into moult. Birds that are being reared for meat and surplus males will start to be ready for dispatch. Pullets will come into lay. Send off poo samples for worm counts and worm if necessary.
October	Deal with any hut and run maintenance before the weather turns for the worse. Continue to dispatch and dress meat birds, and decide which birds to keep over winter for the following season. As grass quality decreases, feed poultry corn to geese.
November/ December	Geese and turkeys hatched earlier in the year are now ready for dispatch for the holiday season. Continue to manage any rodent infestations.

FEEDING

Most manufacturers suggest feeding birds ad lib; fast-growing youngsters need food available to them most of the time. For adults, give them enough for their daily requirements, and reduce amounts if there's food left at the end of the day as you are almost certainly feeding rodents and wild birds with your generosity. Never leave food or water in the bird huts: overnight the birds will be asleep, and during the day it is best to put food and water in the run to limit dirtying and wetting of bedding. Putting food and water in the hut simply encourages bacteria build-up and vermin.

Keep feed in metal, rodent-proof containers. Poultry feed has a use-by date so make sure that you aren't topping up feed bins with new food on

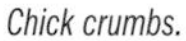

Chick crumbs.

Layers pellets.

Mixed poultry corn.

top of old. The industry standard use-by date is three months from the date of manufacture and bagging, although the vitamins, minerals and trace elements are good for at least four to five months, so there are no performance issues if feed were to go out of date while on your holding.

If you spot any mould in the feed, do not give it to your birds; throw it out, clean your storage bin thoroughly, and source fresh feed. Clean out the feeders too if they get grubby, removing any clogged food remnants before they become damp and mouldy. It might be tempting to buy feed in bulk, but if the number of birds you have doesn't warrant this, you risk the feed going off. Having a month's supply is as much as you want to be storing.

Treadle feeders are particularly useful in deterring rats, and birds can be trained to use them, but they are not suitable for geese.

Poultry and waterfowl feed comes in four age/stage varieties:

- Starter crumbs for newly hatched birds to around 3 weeks of age (19–22 per cent protein).
- Grower/finisher pellets to rear growing birds from 4–10 weeks (16–17 per cent protein).
- Maintenance pellets from 10 weeks to onset of lay (15.5 per cent protein).
- Breeder/layer pellets for birds coming into lay (16–17 per cent protein).

The lower protein range is for chickens, higher for ducks, while turkeys and quails need even higher protein levels.

Treadle feeder suitable for hens and ducks.

Feed Amounts for Poultry and Waterfowl

Species	Feed (approximate daily quantities given per adult bird)
Chickens	Large fowl eat about 110-170g of pellets or mash (whole and partially ground grain) per day. Bantams need about 50-85g depending on size.
Ducks	A large table duck will eat quantities of feed (growers or layers pellets according to age), whilst a runner duck has a modest appetite on a par with chickens. Estimate 150-200g per adult.
Geese	Geese should have access to grass for the majority of their diet; they can be on 100 per cent grass during spring and summer. When the grass gets poorer give 125-200g mixed poultry corn or wheat (no more than 150g for every 5kg bodyweight).
Turkeys	25-26 per cent protein for baby poults (starter crumb), 21-23 per cent protein for growing birds, 15-18 per cent protein for finishing birds to meat weight or as adults. 130g (hen), 225g (stag) and less if grazing.
Guinea Fowl	Guinea fowl graze and enjoy greenery, and are commonly fed on chicken feed, 100-150g (growers/layers etc) depending on age.
Quail	Quails eat 15-20g of quail feed per day, at high protein levels of 27 per cent when growing, and 19.5-20 per cent as adults by 50 days.

A range of suitable feeders.

Geese eating mixed poultry corn from a trug.

Feeding Chickens

Chickens love fruit, vegetables, weeds, mealworms, trespassing insects, frogs, mice and more, but compound feeds (layers pellets and the like) are formulated to provide the precise nutrients your birds need, so don't overdo the treats. You may prefer to give half the ration in the morning and half in the afternoon if that suits your routine and minimises loss to wild birds and vermin; however, you can put a day's ration in the feeder and serve it up at one go, as the birds don't gorge it all at once. In cold weather feed consumption may need to go up, so adjust accordingly.

Feeding Geese

Geese are simple to feed, as the vast majority of their diet is grass. This does not mean grass cuttings collected in the grass box of your lawnmower, which will quickly heat up and are only fit for the compost heap. Geese need to graze their own supply of standing grass from a field, paddock or lawn, and will forage for much of the day, consuming over a kilo of grass (equating to 20 per cent of an adult sheep's intake). Because they mostly forage for their own food, allow geese enough time to take in adequate feed, around seven to eight hours.

With the arrival of winter and shorter daylight hours, the quality of the grass will be at its poorest, just at the time when plenty of nutrition is required for the birds to keep warm, so their diet will need supplementing. Wheat or a mixed poultry corn (substantially wheat with a small quantity of maize in the mix, and may also include peas, calcium and grit) should be given throughout the winter. Avoid giving pure maize as this will just make your birds fat, and don't give corn to growing goslings.

To deter squirrels, rats and wild birds from stealing feed, tip it into a shallow bucket containing a few inches of fresh water. Geese are happy to dibble in water to find their wheat, and pests won't have access until the water is drunk or spilled, by which time the feed will have been devoured. You can't do this with poultry pellets or mash, which would turn instantly to mush and be rejected.

When broody geese come off the nest after hatching goslings, they will have eaten practically nothing but a quick daily snatch of feed for a month; to build up their weight, offer them mixed poultry corn for a few weeks. Their goslings will enjoy lettuce, spinach and chard from the veg garden, as well as dandelion leaves to supplement their starter crumbs; adding peas or sweetcorn will give them the added niacin they require.

Feeding Ducks

You can buy waterfowl feed for ducks from hatch through to laying, but it can be expensive and is not available everywhere. If you can source them, feed duck starter crumb for ducklings, which provides a high energy diet for the first three weeks, and if not, use chick crumbs – always the unmedicated kind – plus niacin (*see* Supplements below). Move on to poultry grower or duck grower pellets after six weeks.

Once full grown and coming into lay, ducks should move on to poultry layer pellets, which have a higher calcium content than grower pellets. For those being reared for the table, finisher pellets can be fed. Duck maintenance pellets with lower protein are designed for feeding full grown young birds before they come into lay, as they don't need the additional protein boost of the higher energy breeder pellet until they start laying.

Supplements

Waterfowl require more niacin (nicotinic acid/vitamin B3) in their diet than chickens, particularly in their early growth stages, as deficiency can lead to leg and joint problems. Supplement their diet (via feed or water, depending on the supplement chosen) with liquid niacin in their drinking water, dried brewer's yeast, nutritional yeast flakes, or simply by including sweetcorn or peas in their feed for a niacin boost. From 55 to 70mg of niacin per kilo of feed is recommended, until birds are around ten weeks old. If using nutritional yeast, add one to two tablespoons per 200g of feed.

There are plenty of supplementary feedstuffs that can be fed in moderation: bolted lettuces, windfall apples, sprout tops, past-their-best corn cobs and other greenery from the veg patch and orchard will be appreciated. Give handfuls rather than bucketfuls or barrowloads to your birds, because it will just sit and rot and be yet another reason for vermin to visit. Many people give a regular afternoon 'cream tea' equivalent of maize to their birds, which is unnecessary and fattening, but portions of apple, lettuce, cucumber or other greenery will keep birds busy and happy without being gut-busting. For those of you who intend to dispatch and dress your own birds, you'll soon see the more-than-adequate natural fat deposits, and you don't want to create excess quantities of fat by overfeeding with maize.

Grit

Sand and grit enable the bird's gizzard to grind up food effectively, so have sand and poultry grit available so birds can help themselves. Birds that free range probably get enough natural grit, but if they are in pens or if you have concerns about their ability to digest their food, or if you are finding soft-shelled eggs, do give them access to poultry grit: this is normally flint grit with oyster and other sea-shells, which also offer calcium. Keep this separate from their feed so they can help themselves as necessary.

Drinking Water

All birds need clean drinking water available throughout the day. Chickens manage happily with galvanised metal or plastic poultry drinkers. The

Shallow bucket – a cheap, cheerful and effective option for drinking water.

Chick drinkers – also suitable for very young waterfowl.

Galvanised drinker for growing birds.

Automatic duck drinker.

easiest and cheapest option for adult waterfowl is to use horse-feed type, shallow, low-sided 15ltr buckets with handles that can be taken to a tap, if a hose can't be taken to them. You will need at least one of these per pair of geese, and per trio of ducks. The shorter the bird, the shallower the water container needs to be, and a clean, new, cat-litter tray (without the litter, naturally) is a good option for birds of short stature.

When the weather is hot you'll need to refill drinkers at least once, if not more often during the day, or increase the size of, or double the number of drinkers.

For gut health you can add some organic cider vinegar to the water (2ml cider vinegar to a litre of water given in a plastic drinker), but always offer untreated water alongside.

HOUSING

There are so many types of poultry housing available – wood, plastic, rubber, posh, expensive, workmanlike, DIY and so on. A poultry house can be purpose made, or repurposed from any sort of small building such as a garden shed, stable, pig ark, goat hut, dog kennel or lean-to. If you want new, there's nothing wrong with buying a garden shed for your birds – it will be cheaper than a new bird hut, and tall enough for you to move around inside. Helpful adjustments include lining the base with weldmesh to deter rats, and replacing glass windows with acrylic sheet and/or weldmesh, depending on the prevailing weather and the amount of ventilation needed.

Most important is to choose something robust that a fox or badger can't chew through overnight, and that is easy to clean. Ventilation is critical, but avoid creating draughts.

In damp areas (which will be much of the UK), avoid softwoods for housing if you want a long-lasting hut – although they will do the job for a few years – and avoid shiplap as it offers too many crevices for mites. Recycled plastic huts are very long-lasting, easy to clean and less prone to mite infestations.

How much space do birds require? Large fowl chickens and large ducks require approximately 0.4sq m (4.3sq ft) per bird in the hut if birds are also able to roam during the day. Bantams and small ducks can cope with half this space. Geese will need

Mobile chicken arks.

Recycled plastic chicken coop.

Covered poultry run.

Covered chicken run and hut.

Duck house with ramp.

about 1sq m (10.76sq ft) each. Most breeds of chicken will appreciate a perch, 5–7.5cm (2–3in) wide depending on the size of the bird, with rounded edges; 25–30cm (10–12in) of perching space per bird should be adequate.

Encourage egg-laying with one nest box for every four or five chickens. Nest boxes should be raised off the ground at least a few inches, but lower than the lowest perch, while ducks and geese simply create their own nest in the straw bedding.

To deter rats from nesting under static huts consider putting down a thick concrete pad as your hut base, which could be made with an additional apron large enough to stand drinkers, to avoid muddy bogs round the drinking area.

Ensure doorways are wide enough so that birds don't tumble over one another and squeeze out uncomfortably were the opening too narrow, potentially damaging themselves and each other. If a hut

Poultry huts and cage runs with sheets for shade.

Modular run and hut.

Hen with chicks.

Mixed eggs in the incubator.

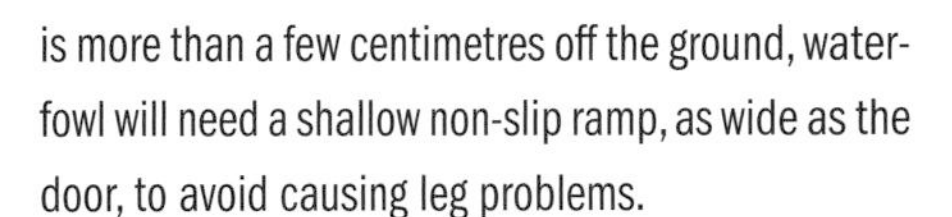

is more than a few centimetres off the ground, waterfowl will need a shallow non-slip ramp, as wide as the door, to avoid causing leg problems.

Create an area that throws shade to protect your birds in very hot weather – this might be an extended roof with open sides, or situate the house and run under the protection of trees, shrubs or hedging. For cage runs, cover with old sheets on a hot day to create shade, or move them into areas where there is natural protection. However, ensure you don't have overhanging branches that are accessible to foxes, which can use these as a way into your pens. In icy weather put a solid-topped pallet or a heap of straw on the ground so birds can get off frozen soil and snow.

Types of Bedding

From the tender age of a day old to maturity, ducks and geese produce copious amounts of muck; chicks are less mucky. All newly hatched birds can have shavings as bedding, while for growing and adult birds use straw, softwood shavings or shredded paper. All these compost well and are a great addition to the veg bed once well rotted. Avoid hardwood shavings as they contain tannins that are toxic for the birds. Hay is unsuitable as bedding as it can get tangled round their feet and becomes a very heavy, moisture-retaining mat, beloved of bacteria and mould. Avoid sawdust as birds can eat or inhale it.

How often you should muck out bird huts doesn't have a timetable: it will depend on the size of the hut, the number of birds, the weather and the type of bedding. Make time to do your mucking out weekly if that works for you, but you may find in good weather that a cursory picking out and a scattering of additional straw is all that's required, while in wet weather a thorough going through every few days with a generous layer of fresh bedding is needed. When birds emerge from their huts each morning, they should be cleaner and drier than when they went to bed.

LAYING, INCUBATING AND HATCHING

Not all eggs are fertile, and female birds will only produce fertile eggs if they are running with a male that breeds with them; nevertheless they will still provide wonderful eggs for the table if in an all-female group.

It's easier hatching and rearing chicks under a bird, but this is not always possible or advisable, particularly if you struggle to provide a rat-free environment. Some birds don't go broody, but you can use a broody hen to sit on eggs of any species, and they will happily sit the extra days required for duck and goose eggs. It is not possible to buy eggs that are a few days away from hatching; removing them from the incubator so they can be posted elsewhere, or

Incubation Times

Species	Incubation period (average)
Chicken	21 days
Duck	28 days
Goose	30 days
Turkey	28 days
Quail	18 days
Guinea fowl	28 days
Muscovy duck	35 days

transported to another incubator hours away, will kill the developing bird. No breeder would agree to sell late-stage incubated eggs, although they'll be asked to do this with surprising frequency.

Nature is not always precise, and whether hatching artificially or naturally, depending on the temperature, humidity and weather, incubation can take a couple of days less or more than the incubation times given above.

Breeding Naturally

Timings for coming into lay vary depending on location, weather, breed and the vagaries of nature, but by the end of February or early March things do tend to get going for most birds. The first eggs are often laid days apart, but once they've come properly into lay, birds will produce an egg every day or every other day more or less, depending on species and breed. As the days go by and the clutch of eggs increases in number, the female starts to contemplate sitting, and plucks down from her breast to line the nest. Although not uncommon, it's not ideal for birds to share a nest: the chances of them stealing each other's eggs are high, and eggs are overly disturbed and can get cold.

Birds often create a clutch of eggs larger than they can effectively sit on, so a certain amount of human intervention is required. When the female starts to sit solidly, remove the dirty, oldest eggs and leave her a number she can spread herself over. A duck can sit on eight to fourteen eggs, although a big Muscovy may sit on as many as twenty, a goose on eight to twelve eggs, and a hen on ten to twelve eggs of her own kind. Remove surplus eggs that go beyond a number that your bird can manage when she hops off the nest for a quick feed, to relieve herself and drink some water. A small broody hen could theoretically manage two or three goose eggs, although you may have to turn the eggs manually if she finds this challenging, while a large breed hen will cope with five or six goose eggs depending on her size, with the considerable advantage of being able to turn the eggs herself.

Place fresh water and feed close to the nesting bird (not inside the hut, but close enough to it so she can keep watch over her eggs as she takes her brief break).

Incubation doesn't start from the day the egg was laid, but when the bird starts to sit firmly on her clutch of eggs, so that all hatchlings appear over a couple of days. Around two days before hatching is due you can hear cheeping from the eggs. Remove any large water containers so the chicks can't drown, and put out a suitable drinker so that they and the mother can drink. The mother can happily eat chick or duck starter crumbs put out for the baby birds, as well as adult feed: raise this off the ground so that only she can reach it. If you allow birds to hatch naturally, the hut where the mum and her young are nesting and any attached run must be safe from predation.

Breeding Artificially

Is there a bird-related topic that is more contentious than that of how to go about artificial incubation? There are huge disagreements as to how this should be done, and just as many individual quirks and approaches for doing something this way or that, but the information below will get you going.

Artificial Breeding Equipment

Equipment	What it Does	
Candler	Artificial light source used to check what's happening inside the egg.	*Candler with egg scope.*
Incubator	A controlled environment that will develop and hatch fertile eggs artificially. Many have automatic turning.	*Incubator.*
Hatcher	A controlled environment that will hatch fertile eggs artificially. It can be used for the whole incubation period, but is often preferred just for the last couple of days for hatching to avoid bacteria build-up in an incubator. It doesn't turn eggs.	*Hatcher.*
Brooder	A brooder is simply a safe area or structure where newly hatched chicks can be fed, watered, kept warm and cared for in the first weeks of life.	*Brooder.*
Heat source	Artificial heat source to keep young birds at the correct temperature while in the brooder.	*Heat lamps.*

When collecting eggs for incubation, make sure the birds have plenty of clean bedding to lay in. Eggs should be clean, well-shaped, free of dings and cracks, and of average size for the breed. Avoid anything that's oddly shaped, wrinkled, miniature or huge. At all times from collection to hatching, eggs should be handled gently. If eggs are slightly grubby, gently brush off any dirt with a clean scouring pad; if very dirty, don't use them for hatching. Bacteria will penetrate a very dirty shell and will multiply in the warm environs of the incubator, causing eggs to go rotten and explode.

Put the eggs in the incubator at the same time, not individually as they are laid. You can incubate eggs up to ten days old. Do not expect anything like a 100 per cent hatching rate; success will depend on so many issues, including the unpredictability of the breed (some are notoriously difficult to hatch, and others have very high success rates).

Candling

Candling is using artificial light to check what's happening inside the egg, and is an important process for successful incubation. It enables you to check eggs for any cracks or defects before you put them into the incubator, to remove non-fertile eggs from five to seven days, to see if fertile eggs are developing as they should, and to identify any that were fertile but are no longer viable.

Weekly checking for and removing unviable eggs is critical, as the bacteria they give off can kill viable eggs, and create a vile mess and stink in your incubator if they explode. Candlers are not expensive, and you can use a makeshift version with a small torch and a cardboard tube from inside a toilet roll if doing a one-off hatching. The more powerful candlers are helpful for eggs with thick shells, and are further improved with the addition of an egg scope that shuts off surrounding light.

The Incubation Process

Incubators have become increasingly complex, with still-air or fan-assisted (forced air) options, manual and automatic turning, programmable automatic humidity control, digital displays and alarms. Your spotlessly clean and disinfected incubator should be brought up to 37.2–37.4°C, and then you have two choices: wet or dry incubation. Follow the instructions that come with your incubator, as every make has its quirks. I am a convert to dry incubation, and don't put any water in the incubator channels at all until the eggs start pipping.

If you want to mark them in any way (noting the breed, the specific bird that laid them or the date for example), write on the eggs in pencil. Don't cram the eggs in too tightly in the incubator as that might crack the shells, and check that the incubator is working each day; a mains failure alarm is invaluable.

At five to seven days candling will show a little nucleus with red veins radiating outwards. If you have any clear eggs, throw them out now. Check again in another week, and again in week three. If you see a red ring or big black spot it means that the eggs have stopped developing or have died in shell; get rid of those, too. Egg weighing is also a good way of tracking development: most eggs need to lose 13 to 15 per cent of their weight by hatching time.

High humidity in the hatching stage is needed to prevent the membranes drying too fast and becoming tough and difficult to tear as the bird hatches. Two days before hatching is due, stop turning the eggs and remove any dividers. Make sure the water channels are full. To add a contradictory note, I've hatched successfully in entirely dry incubators more than once. Do ensure adequate ventilation at this stage – check your incubator instructions – or you may get late death in shell.

Two days before hatching is due and before the eggs start to pip, I move them from their forced-air incubator to a still-air incubator used solely as a hatcher, which can take a lot of water in the base; fan-driven incubators used for the main period of incubation can dry things out too much at hatching time.

Some birds are out of their shell and bobbing about in twelve hours, some take much longer, up

Progress inside Candled Eggs

Candled after 5–7 days: showing clear means the egg is infertile.	*Clear egg.*
Candled after 5–7 days: showing a blood spot and radiating blood vessels is a fertile egg.	*Fertile egg.*
Candled after 5–7 days (or later during incubation): showing a red ring or black spot means early death – the egg is no longer viable.	*Black spot.* *Red ring.*
Egg showing sloppy, mobile, decomposing contents means that it is no longer viable.	*Decomposing egg.*
After 12–14 days the content of viable eggs gets darker and the air sac more defined. The content may now be too dark to see much detail.	*Developing egg.*
Depending on the species, at between 16 and 21 days the air sac is bigger and the embryo contents even darker.	*Further development.*
2 days before hatching, you can see the bill in the air sack.	*Close to hatching.*

to forty-eight hours from first pip. If the membrane is drying out and sticking hard to a partially hatched bird, intervene by carefully picking the membrane off the hatchling, or it might suffocate. Stop immediately if this causes any bleeding. But apart from this, and removing any hatchlings that are 12–24 hours old in batches, don't open the incubator unless absolutely necessary. If you do, you get rid of all that nice humidity you have worked so hard to create. Leave a minimum of six hours between openings, and twelve is better.

Brooding

From the incubator/hatcher put the hatchlings in a brooder (either home-made, or smart shop-bought, or perhaps an adapted hamster cage or rabbit hutch) somewhere safe – away from cats, dogs, magpies and rats. It is important that the brooder has a non-slip surface to avoid the newly hatched birds developing splayed or spraddled legs – this might be an old tea towel, thick textured paper or non-slip mats topped with softwood shavings.

Chick in the brooder.

Safety brooder.

Hatchlings should be under heat for two to three weeks depending on the weather – if the weather is warm you can get away without heat during the day towards the end of that period. Start the temperature at 32°C (a garden thermometer is good for checking this), and reduce this by 1° a day by slightly raising the heat lamp. If the baby birds are clustering together under the heat lamp you know they are too cold; if they are spread to the furthest corners of the brooder they are too hot; if they are acting normally – feeding, drinking, moving about, interacting or snoozing – they are warm enough. Use a ceramic bulb, also known as a heat emitter, rather than an infrared bulb as a heat lamp so that it's dark at night.

Always make sure the brooder is large enough to allow the birds to get away from the heat source, and don't overcrowd it. An excellent alternative to a heat lamp is an 'electric hen' or safety brooder, raised to allow the birds underneath when they choose.

REARING

Newly hatched birds feed off their internal yolk sac for the first twenty-four hours. Provide water in a purpose-made drinker, and don't give baby waterfowl a bowl of water to swim in until they start to feather up – they can easily drown as they don't produce their water-repelling oil or weather-proof feathers until they are some weeks older; waterfowl raised naturally take on oils from their mother as she sits and grooms them. Egg boxes make cheap, biodegradable, throwaway food containers. Make sure their drinker is refreshed regularly at least twice a day, and isn't clogged up with feed or shavings, preventing access to the water. At four weeks old, now off heat, they can go outside in a predator-proof pen and hut.

SEXING BIRDS

Vent sexing (checking for the penis manually by gently opening the vent of young birds, preferably day olds) is a real skill and tends to be restricted to those raising commercial birds for fear of inexperienced hands unwittingly causing damage. There are various alternatives for checking the sex of young birds, depending on the breed, but by the time they are adult even the most inexperienced keeper should be able to tell a cockerel from a hen: the comb and wattles tends to be brighter, the spurs longer, their hackle, saddle and tail feathers are softer, more vibrant, longer and distinctive (in many breeds the cockerel is more spectacular than the female), and of course they begin to crow. Most of these secondary sex characteristics become visible around eight weeks of age.

Some birds are autosexing, their colour indicating at hatching whether they are male or female. The Cream Legbar and Plymouth Rock are autosexing chicken breeds, the female having a dark stripe along her back, the males being paler with a white dot behind the head, but unless you have several chicks to look at and compare you might still get the gender wrong. Commercial hatcheries produce sex-linked birds, chicks that are first-generation hybrids of two separate chicken breeds and have visible differences between male and female; however, this does not continue down the generations if interbred, unlike autosexing birds.

The colour of the day-old baby fluff in some breeds of gosling (and the colouring of their feathering as adults) can be used to distinguish the gender in autosexing breeds. Autosexing geese include Pilgrim, Shetland, West of England and Normandy breeds. Otherwise, telling the sex of geese can be challenging until they are several months of age when their behaviour starts to give clues.

Many duck breeds have significant colour and pattern differences between the male and female, making them easy to tell apart, just as the wild mallard where the showy male has an emerald head, while the female is speckled brown and less vibrantly coloured. This isn't any help for all-white birds such as the Pekin or Aylesbury. In some breeds (for example the Rouen, Appleyard and Welsh Harlequin) the duck has a dark orange or brown bill, while the male's is green – but this is not a rule of thumb, as in some other breeds the females have green bills as well as the males.

Most ducks, apart from the Muscovy, can be sexed by voice by the time they are fully feathered at five or more weeks. The female will give the ubiquitous 'quack' expected of ducks, while the male emits a raspier sound that tends to be more insistent. To be sure that the sound you are hearing is from a specific bird, pick it up, as handling often encourages a bird to quack or rasp. The Muscovy male is around sixteen weeks old before you can sex by voice, when he starts to hiss. At sixteen to twenty weeks drakes will develop distinctive curled tail feathers, apart from the Muscovy, which doesn't sport a curled tail.

Cockerels, drakes and ganders are larger than their female counterparts, sometimes significantly so, as in the case of the Muscovy, where the male can be twice the size of the female, and has larger caruncles around the face. For all breeds, the males tend to have larger feet and thicker legs than the females. Adult ganders have longer, thicker necks and behave more aggressively, but plenty of female geese show protective behaviour, so don't go by that alone. The knobs on African and Chinese geese will develop earlier and become more prominent in the male.

Muscovy male and female mating.

Shetland drake showing the curled tail feather.

Guinea fowl adult males have larger wattles and helmets than females. Turkey stags show obvious differences from the hens once they are adult, from size, to the dramatic tail feathers to their snood, the fleshy growth that dangles over the beak, the pronounced wattle, colourful caruncle and hairy beard, a tuft that grows on the male's breast. As young, it is quite an art to tell the sex of keets and poults.

HANDLING BIRDS

Handle your birds regularly to check that they are of good weight and not showing signs of suffering from infestations of intestinal parasites. Poor catching and handling can easily lead to birds becoming lame; carry them individually, and never catch them by the leg as this brings a high risk of hip dislocation.

To pick up ducks and chickens, place your hands over the wings to prevent flapping, slide one hand under the body, and put the bird's legs between your fingers, positioning one or two fingers between the legs, with the tail facing away from you, the head towards you, and the body lying along your arm. The bird will relax on your arm, and in that position any poo will drop on the ground and not on your clothing. The wings can be controlled by your opposite hand or by holding the bird against your body, under your arm. Ducks in particular have sharp claws so wear gloves when handling.

To handle geese safely, take control of the head to avoid being bitten. Catch the goose from behind by the neck, not applying excessive pressure. Most geese will sit down once caught in this way. Once you have hold of the neck, slide your other hand under the body and firmly clench the legs between

How to carry a goose.

Picking up a large duck.

your outstretched fingers (positioning one or two fingers between the legs), and support the bird's breast on the forearm of the same hand. The head and neck can then be tucked under the armpit of the supporting arm and the bird lifted against your body. The wings can be controlled by your free hand.

Wing Clipping

Most chickens can fly, apart from the heaviest breeds, while domesticated duck breeds don't, with the exception of the Mallard, Call, East Indies and Muscovy. As for geese, most can fly to a greater or lesser extent, although I've only known our Pilgrims fly any distance after sitting and hatching a brood, having lost a significant amount of weight in the process. If your birds are non-flyers, there is no need to wing clip.

Wing clipping removes the tips of the primary wing feather on one wing, so as to unbalance the bird so it can't fly. It's a painless process, like trimming your fingernails or having your hair cut. If you are doing a number of birds, decide to do either all the left wings or all the right, or you may pick up a bird already clipped and do both sides inadvertently. Use sharp scissors to avoid hacking at the feathers. Trimming lasts only as long as the next moult, so you will have to do this annually when feathers have regrown.

Trim the primary flight feathers and avoid the wing tip, which is flesh and bone; it's easy to tell feather from flesh if you feel for the wing tip before you start to cut. You can leave three or four of the large flight feathers at the tip of the wing so it looks more normal when folded against the body. Wait to clip until birds are fully feathered and the feathers have stopped growing and the shafts are no longer filled with blood and are pale or clear in colour. This may be as early as eight weeks for chickens, at fifteen weeks for ducks, and seventeen weeks for geese.

Holding a duck's legs in the proper position for carrying.

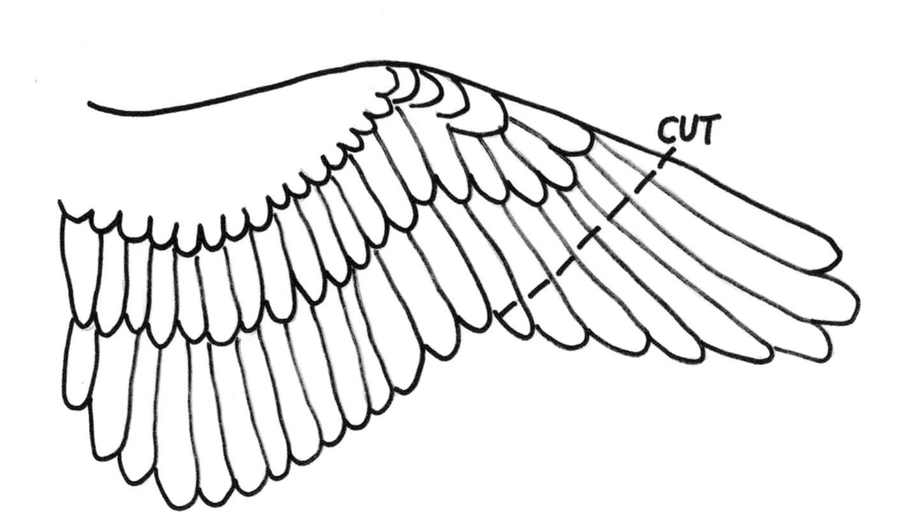

Wing clipping.

MALE-TO-FEMALE RATIO FOR POULTRY

Putting too many males with female birds can cause real suffering and mutilation for the females. With ducks in particular, where the males are the most rapacious, err on the side of caution. On the other hand, you can keep a group of males together very happily, but do not be tempted to put a female in with them 'just for a day or two' – it isn't kind.

Male-to-Female Ratio for Poultry

Species	Number of females per single male
Quail	3–4
Chickens	6
Partridges	1
Pheasants	6–7
Guinea fowl	2–3
Ducks	4–6
Turkeys	10
Geese	4

9 Pigs

Is there anyone who doesn't love piglets? Shiny and new, whiffling investigatory snouts, lively, endearing, snoozing in heaps and as adorable as puppies. All these things are true. What is also true is that even the most pork-loving family couldn't cope with the extraordinary amount of meat that would be produced from the grown litters of an actively breeding sow. A sow may have – and let's not assume an excessive litter size for the purpose of this exercise – ten piglets in one litter. In order to maintain fertility, she'll need to have two litters a year. Estimate that each weaner gives 50kg of butchered meat, multiply that by twenty weaners, and that's a metric tonne (1,000kg) of meat in one year. That's over 15,000 sausages, or 20kg of pork, sausages, bacon, gammon, chorizo, ham and salami per week, every week, with a fortnight off for the Christmas turkey and a rib of beef for New Year. Even the most committed carnivore would be overwhelmed by that tsunami of pork.

If you have a market for your pork products or for the weaners once they reach eight weeks of age, then by all means breed pigs. If you don't, or have no interest in developing a sales outlet for livestock or the resulting mountain range of bangers, then don't breed pigs – simply buy in the number of weaners you need to raise to provide your annual pork supply. Most families will be more than happy with 100kg of pork meat per year, so that's two weaners' worth. At eight weeks old, just over the cusp of piglethood and into weaner status, the pigs will be just as joyful to watch and keep, and will have matured beyond newborn ailments. You won't be overrun with more animals than you can house, won't be worried about keeping a breeding boar, or be gasping at the feed bills incurred by surplus pigs.

SUITABLE BREEDS FOR THE SMALLHOLDER

Lucky are smallholders, to have so much choice from the native and rare breeds that thrive on being kept outdoors. Although some breeds are known particularly as good for pork or great for bacon, all breeds are genuinely suitable for both, with the exception of the Mangalitza, which is best suited for charcuterie as it has a great deal of natural fat and is normally taken to twice the age of other breeds before slaughter. Any of the native breeds below will do you well, as will cross breeds. The weights given are averages.

Suitable Breeds for the Smallholder

Breed	Description (weights are averages)	
Berkshire	Black pig with white markings (socks, tail tip and face blaze), compact, short-legged, dished face and pricked ears. Medium sized, sows weighing 220kg and boars 280kg. Although the bristles are black the skin is pink post-slaughter, so crackling and bacon rind are not black. Revered as the world's best pork by the Japanese because of its rich, flavoursome qualities. The bacon is also excellent. Litter size averages at ten, but fourteen or more is not uncommon. Unlike pink pigs they are not prone to sunburn, and thrive outdoors. Docile and easy to manage. Priority rare breed pig.	*Berkshire.*
British Landrace	Large white pig with heavy drooping ears that originated in Sweden and is now one of the UK's most popular pigs. Well fleshed carcase ideal for either pork (particularly ham) or bacon. Its strength is that it will improve other breeds when cross-bred. Produces large litters of ten to fifteen piglets. Sows weigh 270kg, and boars over 300kg. Priority rare breed pig.	*British Landrace.*

Breed	Description (weights are averages)	
British Lop	Long, lean, large, white lop-eared pig, docile and easy to manage and hardy enough for outdoors. Similar to Landrace and Welsh, but much deeper in the body. Sows weigh 300kg and boars 350kg. Produces well-muscled, lean carcase at pork or bacon weights, good hams and long loins with less tendency to run to fat than other breeds. High killing-out percentage of around 80 per cent. A pork pig (70–75kg) should be ready for slaughter in 160–170 days. Litter sizes twelve to fourteen, and make good milky mothers. Priority rare breed pig.	 *British Lop.*
British Saddleback	Black pig with white stripe or saddle round the body at the shoulders and front legs. Large, hardy pig with forward-facing ears, it likes to graze (don't assume this means no digging). Known as a bacon pig, it is an amalgamation of the old Wessex and Sussex pig breeds. Litters of twelve to fourteen, known as good mothers that farrow easily. Can be greedy so diet needs firm management, and mature boars require vigilance. Sows weigh 270kg and boars 320kg. At risk rare breed pig.	 *British Saddleback.*
Duroc	Originating from the USA, big, hardy red-brown/auburn small-headed pig often crossed with other breeds to decrease the fat in the offspring. Ears are semi-lopped and forward-facing, deep body and broad, full hams. Agile and good mothers with few farrowing problems. Fast-growing, they reach meat weight (and still produce significantly more meat from the carcase than other native breeds) at just five months, achieving a substantial slaughter weight of 100kg. Sows weigh 350kg, boars 400kg or more.	 *Duroc.*
Gloucestershire Old Spot	Large white pigs patterned with black spots, big lop ears reaching the tip of the nose. Gentle, hardy, broad and deep-bodied, known for high quality pork and bacon. Prone to sunburn. Sows make good mothers and weigh 280kg, boars 350kg. Slow growing, diet needs to be monitored to ensure they don't get too fat; aim to reach slaughter weight of about 80kg to produce a carcase 50–60kg. Litter sizes eight to ten. At risk rare breed pig.	 *Gloucestershire Old Spot.*

Breed	Description (weights are averages)	
Hampshire	Large American breed that came from UK stock, but not so popular in the UK. Looks like the Saddleback (black pig with white saddle) but has pricked ears. Offspring can have a wide variation in colour and markings. Used more commonly in commercial systems because of its quick-growing carcase yielding lean meat in generous quantities. Slaughter weights of about 90kg achieved in five months. Active, hardy and feisty animal; litter sizes of twelve to fourteen. Sows weigh 250kg, boars 300kg.	 *Hampshire.*
Kunekune	Small, hairy, placid pig originally kept for meat in New Zealand; in the UK they are often kept as pets. 60–77cm tall, the adults weigh between 60 and 100kg. Come in a range of colours including cream, ginger, brown, black and spotted. Medium to short snout, pricked or lop ears, short legs and short, round body. Most have a pair of tassels (piri piri) under the chin. Known as grazers, some dig more than others. Thrive on a diet of grass, fresh fruit and vegetables, and do not need high levels of protein: this should be adhered to as they have a tendency to be fat. Reared on grass and ready for slaughter at 9–12 months. Litter size six to eight piglets.	 *Kunekune.*
Large Black	The UK's only all-black pig, large, docile and hardy with lop ears, good for pork and bacon, fast maturing and not one of the greedier breeds. Sows weigh 300kg, boars 350kg. Porkers normally raised to 80kg, producing a 60kg carcase. As with the Berkshire they kill out white so no black rind or crackling. Happy to graze, but this does not mean that they won't dig – they will. Litter sizes nine to ten piglets. Priority rare breed pig.	 *Large Black.*
Large White	Large, lean white breed with distinctive prick ears. Sows weigh around 260–300kg, and boars 350–380kg. Average litter eleven to twelve. Popular in commercial systems; friendly, hardy, prolific, milky, and have a high daily weight gain. Predominantly known for bacon; they have good conformation and little fat. A keen digger. At risk rare breed pig.	 *Large White.*

Breed	Description (weights are averages)	
Mangalitza	Long curly coat, docile, active and generally with lop ears, traditionally found in Austria, Germany, Hungary, Romania and Switzerland. Known as a lard pig, with dark well-marbled meat and high levels of fat that does not go rancid, making it suitable for charcuterie. Good mothers, produce small litters of around six but possible up to eleven. Colours are swallow-belly (black with white belly), red and blonde. Piglets are born stripy, like wild boarlets. Sows weigh 170–180kg, and boars up to 220kg.	 *Mangalitza.*
Middle White	Hardy, medium-sized, stocky and compact white pig with distinctive snub nose and large prick ears; requires good shelter as can struggle with extreme heat and cold. Average litter size is nine piglets. Good maternal instincts and a gentle nature. Roots less than other breeds. Sows weigh 200kg and boars 280kg. Early-maturing pork pig reaching 65–70kg liveweight at 3–4 months, with good killing-out percentage. Not so suitable for rearing beyond that age as tends to put on fat, but makes a popular suckling pig at 10–14kg liveweight. Priority rare breed pig.	 *Middle White.*
Oxford Sandy and Black	Medium to large coloured pig known as the 'plum pudding' pig, ranging from pale sandy to darker rust with black patches, pale feet and lop ears. Natural forager and browser with good placid temperament, including at farrowing. Finishes more quickly than many traditional breeds, and less likely to run to fat, producing excellent bacon and ham. Sows weigh 250kg and boars 300kg. Average litter size is ten to twelve piglets. At risk rare breed pig.	 *Oxford Sandy and Black.*
Pietrain	Medium size Belgian white pig with black spots (piebald) and erect ears, between pricked and lopped. Alert and active, yet docile. Double-muscled, which accounts for its very high and lean meat yield, producing carcases with the highest meat content of any breed. When crossed with other breeds can reduce the fat of the offspring by 20 per cent. The gene for double-muscling is linked to porcine stress syndrome, which can result in sudden death from minor stress. Blood tests can assess the risk, and stock without the gene can be bought from good breeders. Sows weigh 220–240kg, boars 240–260kg. Litter size up to thirteen piglets.	 *Pietrain.*

Breed	Description (weights are averages)	
Tamworth	Very hardy, ginger coated, prick-eared, long-snouted and long-legged, lean pig. Known for being alert, friendly yet protective. Particularly resistant to sunburn. The breed is least improved and unchanged by imported pigs. Slow maturing, good for pork and bacon (originally known as a bacon pig). Sows are milky and weigh 260kg, boars 320kg. Aim for 80kg slaughter weight to produce a 50–60kg carcase. Average litter size is seven to eight. Priority rare breed pig.	*Tamworth.*
Welsh	Friendly, long, lean, lop-eared white breed with well-developed hams that does well outdoors. The most commercially developed of the traditional breeds, reaching meat weight quickly. Average litter size eleven piglets. Sows make good mothers and weigh 150–200kg, boars 250kg. Ideal for pork and bacon production, with carcases remaining lean at heavier weights. Expect a 55–60kg carcase from a four-month porker, and 90kg from a baconer. Boars make good crossing sires to produce leaner offspring. At risk rare breed pig.	*Welsh.*

BUYING IN WEANERS

Most smallholder pig keepers start (and often continue) with buying in healthy, manageably sized weaners at around eight to ten weeks old to grow on for pork and bacon. There are very real advantages to this: you get all the fun of rearing your pigs, seeing them grow, giving them a good diet and deciding how you want them butchered to your particular preference, and then have a fallow paddock left to regrow and regenerate whilst you start munching through your freezer full of pork and gammon joints, bacon and sausages. This also avoids the cost of feeding, handling and housing a breeding sow or boar.

Berkshires weaned at eight weeks.

You can experiment with different breeds, and buy in just as many weaners as you need, accepting that you must have a minimum of two for company and mutual warmth. If friends and family love your roast pork and want to buy a bit, then increase next year's number of purchased weaners. Four or six weaners are a very different proposition from twenty, year in, year out, not just in terms of meat quantity but also the number of arks you'll need, the number of pig paddocks, quantity of fencing, watering arrangements and feed bills.

Buying in boar weaners is the best choice for first-time pig keepers, as it's highly unlikely that you'll want a full-grown boar on your hands, so you are far more likely to follow through with taking them to the abattoir. Boar taint (an unpleasant smell and taste that can come from mature, uncastrated males) is uncommon in rare breeds, and only found in 10 per cent of commercial boars. We have taken countless rare breed boars up to eight or more months of age for over thirty years, and have never had an incident of boar taint, even though the occasional unhelpful slaughterman will curl up their nose when they see these immature boars with their gonads still intact. Routine castration for pigs is not carried out in the UK, and there is simply no need for it.

The additional benefit of choosing boar weaners is that they will grow larger than their female litter mates, will produce more meat, be just as friendly in the paddock, and will be too young at slaughter age to have developed tusks. Buying gilts (young females that have not yet had a litter) of meat quality and then deciding you'll breed from them is not advisable; only the best females should be bred from.

Check with the breeder if the weaners have been wormed; if not, do worm them when you get them home. If possible, keep them inside (in a barn, shed, or even a sizeable livestock trailer) on straw for a few days to make sure none are coughing, are itchy with lice, or have diarrhoea so that you can medicate them if necessary. Only then turn them outside into a warm, straw-filled ark, the entrance facing out of the wind.

It's easy to overdo the feeding and cause stomach upsets, so ask the breeder how much the weaners are eating, and whether that's in two or three meals a day. Don't be surprised if they eat less at first due to the stress of relocation and the 'wild' food available in a previously unused paddock.

Don't keep your pigs on the same ground as your grazing animals: keep them contained in dedicated pig areas. They are keen diggers and will trash the sward, and their droppings contain copper, which is toxic to most sheep breeds. Their delight in digging is helpful if you want to clear areas of scrub for a veg patch or other purposes. Kunekunes are not as determined as other breeds to re-landscape their surroundings, but digging is entirely natural pig behaviour. Pigs can be kept on patches of woodland and will enjoy the beech mast, acorns and roots that those provide.

SETTING UP FOR PIG KEEPING

Fencing

Don't even *think* about bringing home your pigs until you have fencing in place (*see* Chapter 15). Believing you'll contain them by putting up a few hurdles or some inadequate perimeter structure in anticipation of putting up proper fencing 'sometime soon' simply won't work.

Gloucestershire Old Spot pigs in woodland.

Water

Those new to pigs find it hard to believe just how playful (aka destructive) they can be. Putting out a washing-up bowl of water for your new weaners and thinking that will do for the rest of the day is a delusion. They will take a suck at the water, chew the bowl, then flip it over and play in the resulting puddle, all within a few moments. Fun as this might be to watch, it isn't giving your pigs the water they need. Even using heavy or weighted-down containers isn't suitable – these will be fouled and give you unnecessary work, as you will have to clean them out and refill them multiple times a day.

Fresh water must be made available at all times and in unlimited quantities to all pigs including newborns, even if they appear to be getting all they need from nursing their dam. Connect drinkers to a permanent water supply, allowing pigs of any age and size to drink whenever they wish to. Bowl or nipple drinkers – note these are liable to freeze in sub-zero

Nipple drinker.

Bowl drinker.

Daily Water Requirement

Weight of pig (kg)	Daily intake (litres)
Nursing piglet	Variable – up to 0.2
Newly weaned	1-1.5
Up to 20kg	1.5-2
20-40kg	2-5
Finishing pigs up to 100kg	5-6
Sows/gilts both prior to service and in-pig	5-8
Sows/gilts in lactation	15-30
Boars	5-8

temperatures – need to be sited at a height within easy reach of all pigs, whatever their age and size. A sheep water trough will need regular cleaning but provides water for longer if pipes freeze. We tend to use a trough for adults, alongside a lower-set bowl drinker for youngsters.

It's important to provide shade and water for a wallow in hot weather for all pigs, but particularly the pale breeds that are prone to sunburn. If you include a standpipe alongside the drinking trough, behind the fence to avoid it being trashed, you can attach a hose and create a wallow as needed on hot days – though be careful never to leave the hose inside the pig paddock if you want to use it more than once. Shade may be well provided by trees and hedges or walls, but a pig ark is not an appropriate source of shade unless it's big enough and placed appropriately to throw a good shadow for pigs to lie in (lying inside an ark for shade is tantamount to you seeking a cool spot in a greenhouse).

If there is a lack of natural shade, create a cheap and simple shade shelter with two long fencing stakes and two shorter ones with a mono-pitch roof made of tin with 100mm insulating board high enough so pigs can't chew it, and no side walls, positioned to cast the best possible shade during the heat of the day.

Housing

Pig arks come in all shapes, sizes and materials, but the main requirement is that they are robust and provide adequate shelter in all weathers and for all ages. Pigs tend to sleep outdoors on warm summer nights, but at all other times will treat their thickly strawed-up arks the same way you do your own bed, snuggling into the straw in colder weather, and when it is warmer lying on top of the bedding, poking their snouts outside the doorway to catch any passing breeze. Traditional curved top arks can be bought as component parts and assembled at home, or readymade, and come with and without flooring. Arks with floors are obviously much heavier than those without and may need several people or a tractor with a front loader to move them, but they do provide additional protection on wetter ground.

Pig snoozing outside a traditional ark.

Sow and piglets cooling off in a wallow.

Pig ark made from recycled plastic.

Pig housing is often found for sale second-hand, in which case give them a thorough clean with an approved farm disinfectant before use to minimise risk from disease and parasites. Homemade is also a good option, though make sure the structure is strong, and that there are no nails and suchlike poking out that could cause injury.

The old-fashioned brick or concrete pig sty with a small, contained walled yard is rarely seen these days, as we all want to encourage natural behaviours in our livestock, preferring a bigger area for pigs to explore and enjoy. However, if your holding has a sty it can prove useful to house pigs in quarantine, as a hospital ward for poorly pigs, for farrowing, and in poor weather to get pigs off sodden or frozen ground to give the paddock time to recover and give the pigs a break from paddling about in the mud. A stable, an area within a larger barn or a solid shed are good indoor alternatives to the traditional pig sty.

It's entirely possible to keep pigs outside at all times, but we have found that having an inside area for pigs is extremely helpful when coping with all the above situations, plus it allows them to drop any mud into a deep bed of straw in the days leading up to slaughter. The abattoir will only accept clean livestock to ensure food safety.

A neat rule of thumb in estimating the amount of ground you need for pigs is that two pigs can be kept on 20 × 20m of land for the five to six months it takes from buying them as weaners to taking them off to the abattoir.

Organic Pig Stocking Rates

Type of Pigs (Weight)	Number of Pigs per Hectare (Ha)/Acre
A sow and her litter (to 7kg)	9ha/3–4 acre
Weaners (7–18kg)	60ha/24 acre
Growers (18–35kg)	28ha/11 acre
Cutters/porkers (35–85kg)	18ha/7 acre
Baconers (over 85kg)	16ha/6–7 acre

TAGGING

Pigs going to slaughter must have a tag/permanent mark from the holding they are travelling from, *not* their place of birth (if different), so you will need to tag the pigs with an ear tag showing your herd number before taking them to the abattoir. There are two ideal times to do this: the easiest is probably when you collect them from the breeder, who may well do this for you if you bring your own tags and applicator.

How to hold and carry a weaner.

At eight weeks old, a weaner can be held in your arms like a baby (one hand holding a rear back leg, your other hand holding the diagonally opposite front leg), and a helper can easily tag an ear while the pig is held firmly. Leave a little growing room as the ear as well as the rest of the pig will get bigger, but avoid tagging at the edge of the ear as a barely attached tag is likely to get caught in all sorts of things, from fences to the exploring teeth of bedfellows.

The other good time for tagging is at least a week before they are due to go to the abattoir. This gives you enough time to have several opportunities to get your pigs tagged, and means that you won't be in a frenzy of concern on slaughter day. It's hard enough for first-time pig keepers to take their much-enjoyed pigs to the abattoir, but even harder if you add the stress of tagging, the probably new experience of loading the pigs into the trailer, and reversing it at the abattoir. A week before slaughter date, at feeding time, scatter the food in a long thin line and have the applicator and plenty of tags to hand. Do *not* catch hold of a pig's ear, as it will automatically pull away from you. Instead, while their heads are down and they are eating, put the applicator with the tag

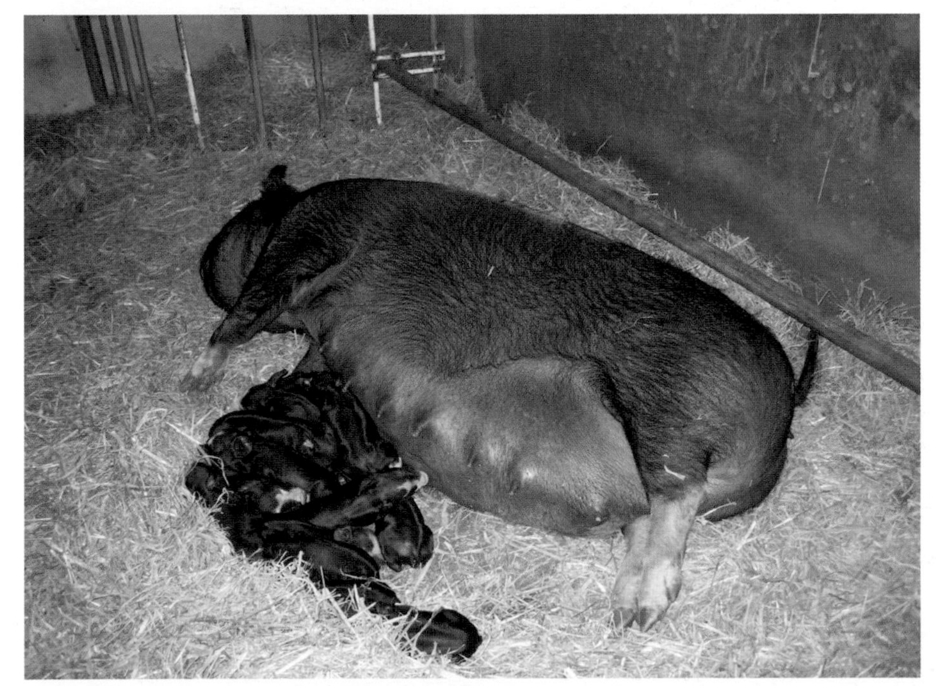

Sow and newborn piglets in an indoor pig pen, showing farrowing bar and creep area.

THE PIG KEEPER'S YEAR

Unlike sheep, pigs come into heat and are ready to mate throughout the year, so a calendar approach to pig keeping is not necessary. However, there are some useful seasonal tips for the new pig keeper:

Spring Source weaners from a reputable breeder. Ensure fencing, water supply and housing are in good order. Buy weaners and bring home.	**Summer** Ensure access to shade and a wallow in hot weather. Adults moult in the summer.
Autumn Weaners bought in the spring will be ready to go to the abattoir in the autumn. Breeding adults develop their winter coat as the temperature drops.	**Winter** In winter you can rest your pig paddock, clean out and disinfect the pig ark, do any remedial work to fencing and check water arrangements, ready for new weaners the following spring.

inserted over an ear and quickly squeeze it shut. Hopefully this will be job done.

Note which ear (left or right) you've tagged, and tag any other pigs in that ear too. This avoids you tagging some pigs in both ears and others in none. If you have a slightly nervous, skippity pig that isn't obliging, you have enough feed times left to tag that one too. I normally manage to tag eight or so pigs in one feeding session; the trick is to be patient and not grab that ear, however tempting that might be.

FEEDING

It's rare for a new pig keeper not to have their first pigs come back from the butcher with inches of unnecessary fat. As much as fat is essential for keeping a pig warm and providing the consumer with flavour, a couple of centimetres enrobing your fine pork chop is beyond adequate, and double that simply tells you that you've been massively overfeeding your pigs, which is not healthy either for them, for you, or your wallet. Talk to the vendor of your weaners and ask how much they are eating. Guidelines on feed quantities from manufacturers are a starting point, but you will need to learn to feed by eye, and to adjust quantities up or down as the pigs' body condition suggests – and never feed more than a weaner can consume within twenty minutes.

Rare breeds use feed more efficiently than commercial pigs so require less protein, and the time of year, the land, and the breed and size of pig will all have an impact on the amounts fed. Weaners are likely to have been on ad lib weaner rations or on three meals, equating to 0.5–1kg each per day. We prefer to go for three meals a day at this age, to minimise surplus food, which will only risk increasing the rat population; we then reduce to two meals within a month.

Observe the weaners at feed times once you have them home, to ensure that all of them are getting enough feed; when feeding at fixed times it is crucial that there is enough trough space for each animal to feed simultaneously to avoid confident animals eating the allowance of their more nervous siblings. Putting feed in a number of well-spaced receptacles can help; the heavy-duty rubber trugs made out of recycled tyres – often used to feed horses – last for years, and because they have handles can be tied to a fence post to stop them being travelled across a muddy, welly-sucking paddock.

Once the pigs are 20–25kg at ten to twelve weeks, give them early grower pellets over two equal feeds a day, moving on to sow nuts, rather than finisher pellets, from around 40kg/sixteen weeks until they go for meat, to avoid them getting overfat. Finisher pellets are more appropriate for commercial pig breeds, rather than for native breed pigs kept on a small scale. When the weather is dry we tend to scatter the feed on the ground; when wet, using a trough minimises the feed turning into slush, but this does mean having to turn the trough over to empty out any mud or rainwater, while the pigs snoofle at your trousers and wellies, eager for their breakfast, so dress accordingly.

Although more costly, it's worth investing in sow rolls rather than sow nuts in wet weather; they consist of the same ingredients but are made into chunkier pieces, which means they are less likely to disintegrate in soggy conditions. For pet pigs, seek out low-calorie pig feeds to avoid pigs becoming obese, with all the related issues.

Protein Levels for Commercial Compound Pig Feed (Note that protein levels vary between manufacturers.)

Pig weaner pellets: 22 per cent around eight weeks (from five to twelve weeks)	
Early grower pellets: 18 per cent (from ten to twelve weeks)	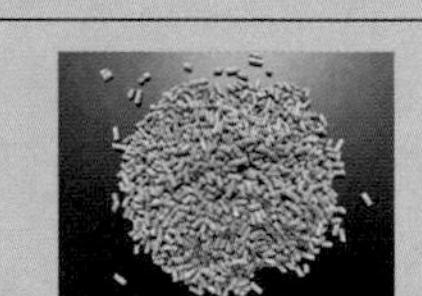
Pig finisher: 19 per cent (from twelve weeks)	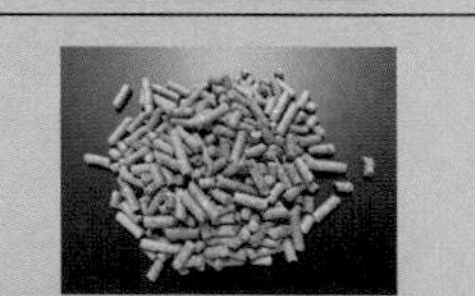
Sow nuts: 16 per cent (from sixteen weeks)	
Sow rolls: 16 per cent	

Liveweight Growth Chart

The following chart gives an approximate liveweight growth and feeding rate for rare breeds of pigs from birth to twenty-four weeks; this guide is suitable for the larger breeds such as the Large Black and Gloucestershire Old Spot, and should be reduced slightly for more moderate sized breeds such as the Berkshire.

Liveweight Growth and Feeding Rate for Rare Breeds of Pigs

Age in weeks	Liveweight range in kilos (approx.)	Daily feed amounts in kilos per pig
2	4-7	Creep feed available ad lib from around ten days - most nutrition comes from the sow's milk
4	7-11	0.22kg or ad lib - still nursing from the sow
6	11-16	0.7kg - still nursing from the sow
8	16-22	1kg now weaned, split into three meals
10	22-23	1kg split into three meals
12	23-30	1.25kg split into two meals
14	30-37	1.5kg split into two meals
16	37-45	1.7kg split into two meals
18*	45-53	1.95kg split into two meals
20	53-57	1.95kg split into two meals
22	57-61	1.95kg split into two meals
24**	61-69	2kg split into two meals
26	69-79	2kg split into two meals
28	79-86	2kg split into two meals

* Feed weaners generously up to four to five months, at which point they are more prone to increase fat levels rather than build muscle if overfed. If they are getting overfat (signs are being jowly round the neck, and barrel-shaped), reduce the feed intake to around 1.5kg, which should finish them without adding surplus fat.

** According to some sources this is the maximum age acceptable as pork for entire boars; however, we take them to seven to eight months (28–32 weeks) and have no problems with boar taint.

THE BREEDING SOW

Gilts (female pigs that have not yet had a litter) reach puberty at around seven months and are mated for the first time between eight to ten months; waiting longer than that may have a negative effect on their ability to conceive. A popular way to start breeding is to buy a sow already in pig (pregnant), which has had other successful litters; alternatively, buying a young gilt of breeding quality will ensure you get to know her as she grows and she will be familiar with your routine and set-up.

Only breed from really good examples. If they are a pedigree animal they must meet the relevant breed standard, and have the accompanying paperwork from the breed society/British Pig Association. Whatever breed or crossbreed, all breeding pigs should have a good body shape with a nice straight topline (a flat back) and not a dipped or a roach (humped) back, they should have at least twelve and ideally fourteen (or even sixteen) evenly spaced parallel teats (both sow and boar), have a good-sized rump, and walk soundly on their tiptoes showing no signs of lameness. Avoid those walking low on their pasterns, or that have uneven or cracked claws on their feet, and the legs should be well shaped (not sickle hocked, post-legged, splay footed, low hocked or pigeon toed). Lastly, go for pigs that don't have fat, jowly necks or are as fat as lard from being overfed.

Females should have a well-developed vulva, and boars should have both testicles descended, and three sets of teats in front of the sheath so that the penis is positioned close to the vagina when the boar mounts the sow.

Gestation

Sows have a neatly memorable gestation period of three months, three weeks and three days (approximately 115 days), although the Berkshire likes to take longer (anything from 118 to 120 days). Healthy sows in good condition may farrow a day or two later than expected, while older sows, first-timers and those in poor condition may farrow a day or two before time. A good sow – this means she has an equable temperament, sound conformation and good mothering instincts – may have eight to ten litters during her life, having two litters a year, although a sow can breed until the age of eight or even ten, though the litter size tends to get smaller as the sow ages.

British Lop sow and piglets.

Mating

A sow comes into season every twenty-one days, although this can vary from eighteen to twenty-four days for individuals, each heat lasting from forty-eight hours in winter to sixty hours in summer, and she will either need to be taken to the boar, a boar brought to her, or be artificially inseminated.

Sows need to be in hard, fit condition at mating (not fat). The most effective time for a sow to be mated is in her first heat period after the weaning of her previous litter (three to five days after removing her from her piglets – we find five days post weaning to be the best time to AI most sows).

Artificial Insemination (AI)

Artificial insemination can be done entirely by the smallholder. Semen is readily available for rare and traditional breeds as well as the modern varieties, and is supplied fresh complete with the required catheters. Your job is to judge when the sow is in standing heat. The vulva reddens and enlarges, and the sow may be more vocal than at other times. Coming into heat does not yet indicate that the female is ready for insemination. Wait another twelve to twenty-four

Artificial insemination kit.

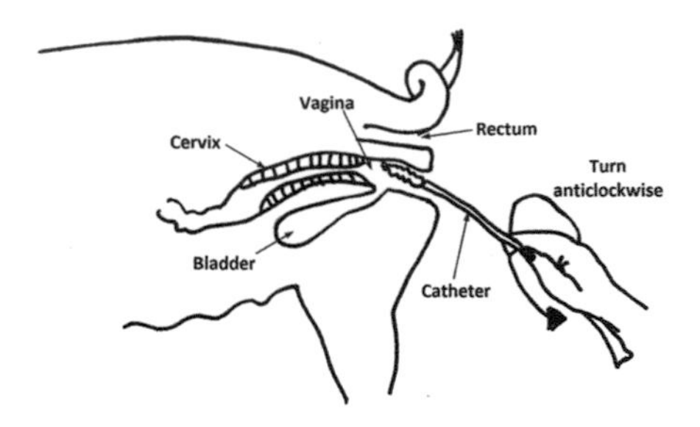

Inserting an AI catheter.

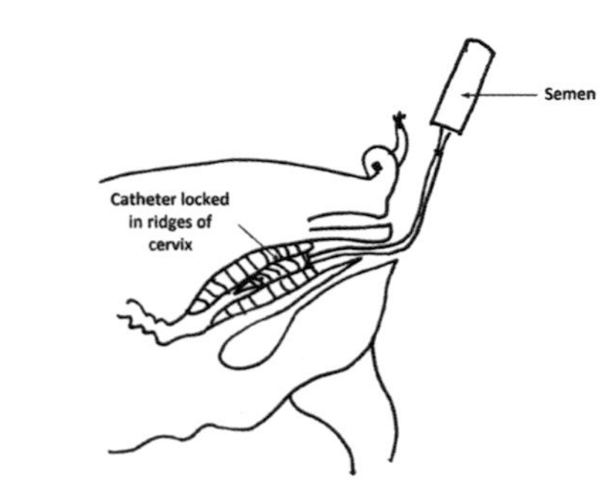

Inserting the semen into the catheter.

hours, checking every few hours to see if she is in standing heat by pushing down on her back: if she moves off she's not ready; when she becomes receptive she will stand solidly when you apply pressure.

Inseminate three times in a twenty-four-hour period (for example 8am, 8pm, 8am), while the sow is standing solidly when you push down on her back. She should not need to be restrained while you do this.

In addition to the semen and catheters you will need obstetric lube (the same used for lambing and kidding), a paper towel, disposable gloves, and you can also use a boar odour spray to further encourage the sow to stand. AI can be done on your own, but we do it two-handed. Ten minutes before insemination I take a bottle of semen and pop it in my bra (a pocket would do!) to warm it to body temperature; once the chill has worn off it's ready to use. The paper towel is to clean off the sow's vulva if necessary, and a line of lube is squirted on to the business end of the catheter. The helper rubs the flanks of the sow, imitating the action of a boar's legs on her sides, throughout the process.

The catheter is inserted gently into the vulva, pointing upwards towards the sow's spine (to avoid the bladder), and turned anti-clockwise until around two-thirds of it is inside the pig. The sow's cervix is ridged, and turning the catheter anticlockwise will lock it into the cervix. If you pull the catheter back gently it should not slide out; if it does, reinsert and continue to turn anticlockwise until it locks in.

Take the bottle of semen, break off the tip, and insert it into the hole in the end of the catheter, holding the end of the catheter upwards, and letting the semen flow down. Do not squeeze the bottle hard: a little gentle pressure is all that is required, and the process should take five to ten minutes, not just a few seconds. Rubbing the sow's flanks helps the semen to be sucked inside.

Once there is no semen left in the bottle or catheter, remove the bottle and then gently twist the catheter clockwise to unlock it, and remove it. You don't want to get a splashback of semen rushing out, so take your time. Don't stress the sow after insemination, or for the next month, in order to maximise a successful pregnancy, and watch carefully from eighteen days onwards to see if she comes back in heat.

Male-to-Female Ratio

How many sows can one boar manage? A mature boar can cope with anything from ten to fifteen sows. The age at which a young boar can first be used to serve a sow varies, but the general rule is that he should be at least seven months. Boars used at too early an age may have their working life considerably shortened and their breeding ability impaired. You may decide to buy a boar of your own if you have enough sows to keep him working, otherwise it is an expensive proposition to feed a boar all year round to serve just one or two sows – and who wants to take on the task of removing tusks from a mature male?

Farrowing

The sow should be wormed a week before farrowing, and checked for lice and treated if necessary. About twelve hours or less before farrowing she will create a nest of her straw (don't give her too much or the new

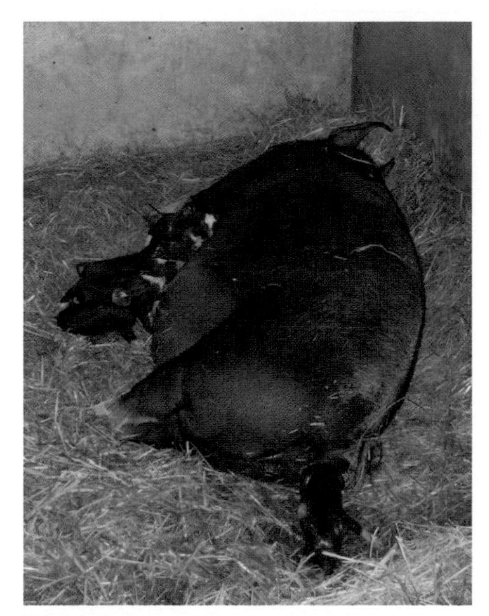

Piglet being born.

Suckling piglets with a creep area.

piglets can struggle to get back into the nest), and her teats will fill with milk. Farrowing can take between one hour and twelve, averaging two to six, and help is rarely needed. Try not to interfere unless absolutely necessary, as squealing piglets can upset the sow greatly and she may become a danger to you and her young. Quietly, so as not to disturb the sow, remove the afterbirth to avoid piglets being drowned in it; she tends to produce some halfway through and the rest at the end.

The sow may have anything from a few up to fourteen or more piglets (we've had seventeen more than once), but be prepared that she may roll on and squash one or more, even if you have farrowing bars in place. Prior to farrowing, create a creep area that the sow cannot enter, where the piglets can sleep under an infra-red heat lamp and later be fed.

For the first two to three days keep the sow's feed light, and then increase rations from the usual amount by an additional 0.5kg per suckling piglet per day across two feeds. Increase this gradually by 0.25 to 0.5kg per day over seven to ten days until she is getting the full amount. In all cases don't feed more than the sow will eat, and feed in accordance with her condition. We find that no matter how many piglets a sow has, she is unlikely to manage more than 4kg of feed per meal.

Feeding the Piglets

Sow's milk is adequate for the piglets for the first two weeks, then you can add piglet creep feed (no more than a cupful to start with), removing anything not eaten or soiled. Feed in a creep area that the sow cannot access. Make sure the fresh water supply is low enough for the piglets to reach; do not use anything the young can drown in.

FROM WEANING TO SLAUGHTER

Weaning

When the piglets are eight weeks old the sow should be removed; make sure they cannot see each other or the sow will get very agitated. At this point we find the sow is keen to be separated from her litter. As soon as she has been removed from her piglets her rations need to be decreased dramatically to get her back into hard condition for mating, and

Piglets feeding.

IRON INJECTIONS FOR PIGLETS

Many resources recommend that piglets born inside should have iron injections, but we have farrowed sows indoors for decades, bringing in the sow a week before farrowing and putting mum and piglets outside at ten days old, depending on the weather, and have never needed to give iron. Newborns source iron from the environment, generally from the soil, but where pigs are kept out of doors for most of the time the sow will have minerals from the earth on her skin, which is enough to provide her young with the necessary iron as they explore around her.

Weaning Poppy the sow – a simple stroll.

to dry up her milk (approximately 1.5–2kg per day over two feeds). Feed the weaners three times a day, observing them to ensure that all are getting enough feed, and treat them with wormer after a couple of days post weaning. They are now ready to go to new homes, or to be grown on for meat, or kept as future breeders.

Ready for Slaughter

A porker is ready to go to slaughter at approximately 80kg, which provides around 50kg of meat. Pigs may reach this weight at anything between four to eight months old, depending on feed, breed, quality of grazing. We take our Berkshires for slaughter at seven to eight months. If you want one pig for pork and the other for bacon and gammon the butcher will automatically use the larger/fatter one as the baconer. If you know the live weight of your pig, expect approximately 75 per cent of that as deadweight, and once fully butchered, a total of approximately 60 per cent of the original live weight in meat, plus the offal, head and trotters, which rounds the yield up to a 65 per cent kill-out percentage.

Weighing crates are excellent but expensive, and your pig may not be keen to get into one. The string method is surprisingly accurate if you can get your pig to stand still long enough, so it's best to do this at feeding time. Measure round the chest just behind the shoulders (in inches) and then measure the length from the ears to the root of the tail. Multiply one measurement by the other, and divide by thirteen, twelve or eleven, depending on whether you think the pig is lean, medium or fat. This will give the pig's weight in pounds, which you can then convert to kilos.

Moving Pigs and Loading them into a Trailer

Moving adult pigs of good disposition isn't much of a challenge. At feeding time rattle their bucket and they will normally follow you, stopping for the odd mouthful of dock leaves or a particularly lush patch of grass, through quite complicated turns, gates and obstacles. Always shut dogs away, and make sure the route is safe. This method allows us to move adults into a barn, out to a paddock or into a trailer.

With youngsters – those being taken to the abattoir, or simply pigs that are a bit skittish – set up an arrangement so that you can feed them for a few days in a trailer or a link box like that shown (the tractor was needed for other chores and was reattached once the weaners were shut in). Curiosity and appetite will get your pigs into the container, but don't expect to do this on day one – allow two or three days for them to feel confident in something with an unfamiliar floor, and where they may have to take a step up.

Preparing for moving young pigs in a link box.

Moving sow and piglets.

Pig Costings: Rearing Weaners to Meat Weight

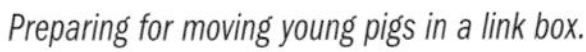

The price of pure native-breed eight-week-old weaners at 2022 prices varied from £40 to £100 each for those intended for meat (and maybe more, depending on availability). For a maiden gilt ready for breeding at around nine months, expect to pay around £200–£450, which can include her being already in-pig. Breeding boars cost £200–£700. Show animals will be at the top of the range.

Rearing your own weaners for meat can provide a cost-effective way of rearing excellent pork; buying pork of the equivalent quality will cost twice as much as rearing your own, if you don't include the cost of the ark, straw, trailer, fencing, water, labour, vet, or the purchase of the land on which to rear them. Hopefully you won't need the vet, and the more expensive of these items should last many years of pig rearing.

Berkshire weaners at feeding time.

10 Sheep

Is there anything more pastoral than a flock of sheep quietly grazing? They don't churn up the sward in the way that pigs do, and are easier to handle. They fertilise the ground as they go, and the sight of lambs hurtling round their pasture in gangs as independence of spirit kicks in, surely gladdens the hardest of hearts.

Pasture-fed sheep kept in appropriately low stocking rates are a marvellously sustainable source of protein, in addition to providing wool and skins. There are so many breeds to choose from, from the robust, oblong, chunky commercial sheep, to the gazelle types of the primitive native breeds: there is undoubtedly something that will appeal to the smallholder in terms of looks, temperament and yield.

HUSBANDRY TASKS OVER THE YEAR

Even if you keep sheep as pets and eco-friendly lawnmowers rather than for breeding and meat production, there are still plenty of tasks to do over the year; it's never the case with livestock that you can simply shut them in a field and let them get on with it as best they may. A quick summary of requirements is given in the table below.

Balwen sheep.

Sheep as Pets versus Breeding Animals

Sheep as pets/lawnmowers	Breeding sheep
Daily observation	*As for pets plus:*
Ensuring they are in good health	Sourcing/keeping ram(s)
Adequate fencing	Lambing
Water supply	Shed/barn if lambing indoors
Grazing management	More winter feed, concentrates for ewes expecting doubles/ trebles
Hay in winter	
Shelter of sorts (hedges will serve)	Constant care over the lambing period
Hurdles and/or basic handling pen	Crutching pre-lambing
Foot care	Worming programme for lambs
Managing internal and external parasites	Vaccination for lambs
Prevention/treatment of fly strike	Ear tagging
Dagging dirty back-ends	Plan for sale/consumption and means to implement
Shearing yearly	Culling policy for breeding ewes
Annual clostridial vaccination	Relationship with abattoir and butcher
Dead stock removal	Space to wean lambs away from dams

Suitable Breeds for the Smallholder

There are over thirty million sheep and lambs and more than sixty pure breeds in the UK, and many more imports and crosses. Not all sheep are the same, whether judged on ease of lambing, prolificacy, temperament, fleece quality, hardiness, meat yield or flavour. As always, what is suitable for one smallholder is not for another: it depends on what you have in mind as their purpose. If chunky meat lambs are what you want then choose from Mules, Suffolks, Dorsets, Charollais, Hampshire Down, Border Leicester, Shropshire or Zwartbles, to name but a few. If the priority or 'at risk' rare breeds are a passion, you might choose from Lincoln Longwool, North Ronaldsay, Whitefaced Woodland or Norfolk Horn (the list of options here is long). And if small breeds appeal,

Hampshire Down.

Cast Whiteface Dartmoor.

Texels.

the Balwen, Ouessant, Hebridean, Portland or Shetland will be suitable.

If you're a crafter, consider the colouring, staple length and handle of the fleece. If the idea of forward-pointing horns is something you wish to avoid, the four-horned Jacob or Manx Loaghtan might not be your first choice. And a novice lamber should avoid box-headed breeds such as the Beltex and Texel. Do your homework before purchasing; some of the primitive breeds, while independent, good doers and easy lambers, can be very flighty, and a new shepherd trying unsuccessfully to contain their flock for husbandry tasks could be put off keeping sheep for good. You can always start with a docile breed and progress to the more primitive types as abilities and confidence grow.

Of our own four breeds, the Torddu and Torwen Badger Faces make excellent first-timer flocks, while the Whiteface Dartmoors require more vigilance (their heavy fleeces mean they are susceptible to being cast on their backs in late pregnancy and in wet weather, and the flystrike risk is higher), while the Herdwicks are for confident shepherds as their determination to keep all four feet on the ground can dismay those still learning how to tip up a sheep for those essential health checks. If you're confused about all the breed options, research those that are native to your region as a solid starting point.

Tipping Sheep

Every sheep keeper needs to know how to tip their sheep on to their bottoms. You'll need to do this for footcare, health checks, feeling udders for lumps/mastitis, checking teeth and any mouth issues, and to immobilise a sheep for whatever reason, including enabling lambs to suckle on recalcitrant mothers. It's imperative that you do not grab, pull or carry sheep by their

Herdwick ewes and lambs.

fleece as this causes bruising and is painful – it is the equivalent of you being pulled around by the hair, and when they are shorn you won't have the option anyway. The technique requires practice but not significant strength, so even someone of slight stature can tip a good-sized ewe.

For large, heavy rams, turning them over is easier with two people. With one person holding the ram in the first position shown, their helper takes hold of the ram's rear leg furthest away from them, and pulls it towards them so that the ram sits on its bottom.

How to Tip a Sheep

Pen your sheep tightly with hurdles so they can't run around, then take hold of the chosen sheep under the chin and by the top of the tail. As a learner you might find it easier to hold it up against a hurdle while you take a breath ready for the next step.	 *Holding the sheep firmly.*
With the hand under the sheep's chin, turn its head away from you so that it is facing along its own spine.	 *Turning the head towards the spine.*
Keeping the head turned towards the spine, immediately with the other hand, fold the ewe's body like a concertina (neck meeting tail) and push the sheep's rump down to the ground. Do not push the sheep between your legs (a common mistake), but down outside your foot.	 *Pushing the sheep's bottom to the ground.*

The Shepherd's Monthly Management Tasks

Month	Management task
January	Feed hay. Scan ewes 80–100 days from tupping date. Start feeding concentrates to ewes expecting doubles/trebles eight weeks before lambing.
February	Feed hay. Feed concentrates to ewes expecting doubles/trebles. Crutch ewes. Vaccinate ewes four weeks before lambing (include lambs from the previous year kept for breeding, and rams too).
March	Clean the lambing shed and set up lambing pens and lambing kit. Bring ewes in for lambing. Send faecal samples for worm counts and treat sheep accordingly/if necessary. Lambing starts.
April	Lambing. Do lambing turn-out tasks, including thorough check of feet and trim if necessary. Flukicide for ewes. Vaccinate all lambs when the youngest is three to four weeks (first dose).
May	Worm lambs from four weeks of age. Ear tagging. Second vaccination of lambs 4–6 weeks after the first dose. Stay alert for pink eye and orf outbreaks. Hoggets (last year's lambs) to butcher.
June	Worm lambs. Shear ewes and rams. Spray lambs against flystrike. Dag lambs if necessary. Haymaking.
July	Haymaking.
August	Wean lambs at 4–5 months old. Pick out the best ewe lambs to keep as future breeders. Put lambs on best grass to fatten/sell surplus. Put the ewes on poor grazing to dry up their milk.
September	Pre-mating check: teeth, tits and toes for ewes. Keep those fit for breeding, the rest go for mutton. Teeth, testicles and toes check for rams. Flukicide for all sheep (apart from those going for meat). Lambs to butcher. Overwintering lambs shorn.
October	Put teaser ram with ewes on the 1st of the month. Put breeding ram(s) in with the ewes on 15th October. Meat lambs to butcher.
November	Start to feed hay. Remove rams after two cycles (35 days).
December	Feed hay. Unsold lambs not being kept as breeders should be maintained over winter for meat as hoggets.

Once their bottom is on the ground, immediately lift the sheep's front legs up.	 *Lifting the front legs away from the ground.*
You now have the sheep seated (shift it on to one 'buttock' or the other, rather than on its tailbone, for its own comfort). When you become accomplished, these stages all combine into one quick movement.	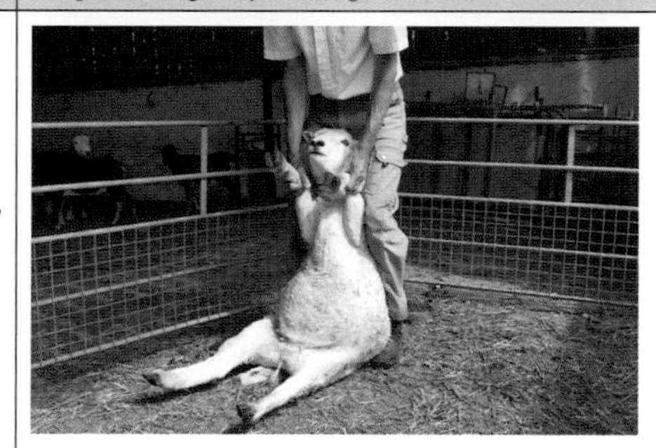 *The sheep seated.*

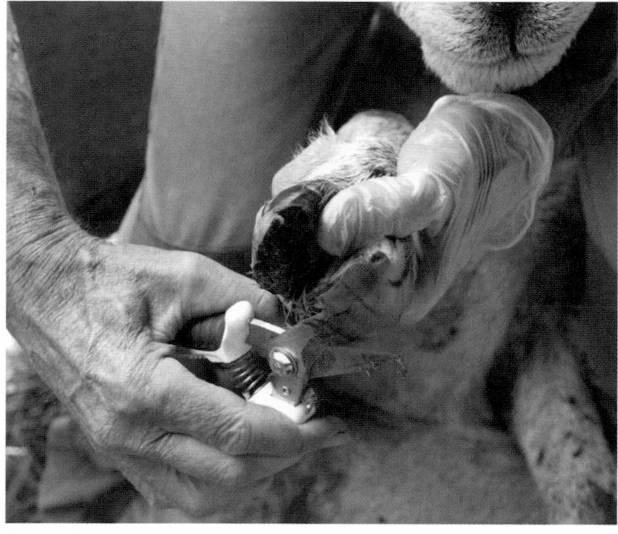

Foot trimming.

The Shepherd's Year

The timing of events depends on what part of the country you're in, the breeds you have, and when you lamb, if you choose to breed. The further north you are, the later things will happen. The panel refers to a typical year for us in Devon; adjust for your own location and system. If you don't breed, some stages will not be necessary.

BREEDING PREPARATIONS

Our sheep year starts in early autumn when we do all the pre-breeding preparation.

Pre-Tupping Checks

The run-up to autumn is the time to get sheep in tip-top shape for tupping (mating), and to check the health of teeth, udders and feet. Successful tupping is followed by five months of pregnancy for the ewes, when you handle them as little as possible to avoid stress (theirs, not yours).

Health management tasks should be carried out at least four weeks before you plan to introduce the ram. Sheep should be wormed and fluked if necessary (we rarely worm adult sheep, but drenching with a flukicide is necessary in wet areas: *see* Chapter 13), and any foot trimming should happen now, but only if needed. Some shepherds crutch their ewes, shearing the fleece from the ewes' tails and back ends to aid mating, but I've never known a ram to be deterred by a fully bushed tail.

Ewes should be checked for a good udder (no hard lumps indicating mastitis), and for their body condition. If a sheep is too thin it needs to be fed to put on condition before going to the ram, while overly fat sheep are at risk of pregnancy toxaemia later in the season, which puts the ewe and lambs at risk.

Check the teeth to make sure they are in good order, as this affects a sheep's ability to get the nutrition it needs. Only fit, healthy ewes should be considered for mating. Problem ewes – those that are too thin, broken mouthed, have a poor udder, a history of lambing problems or persistent lameness – should be identified and culled as mutton.

TEETH DEVELOPMENT

The average life expectancy of a sheep is ten to twelve years, though some sheep may live as long as twenty years. Teeth tell the age of sheep, goats and cattle – but note that at the front of the mouth on the upper jaw they have a dental pad and no teeth, whatever their age. At the side of the mouth they have upper and lower molars. (*See* Chapter 7.)

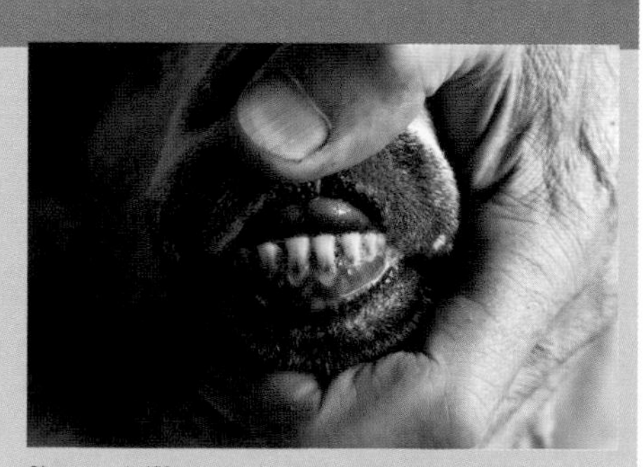

Sheep teeth (fifteen months old), showing the upper gum pad.

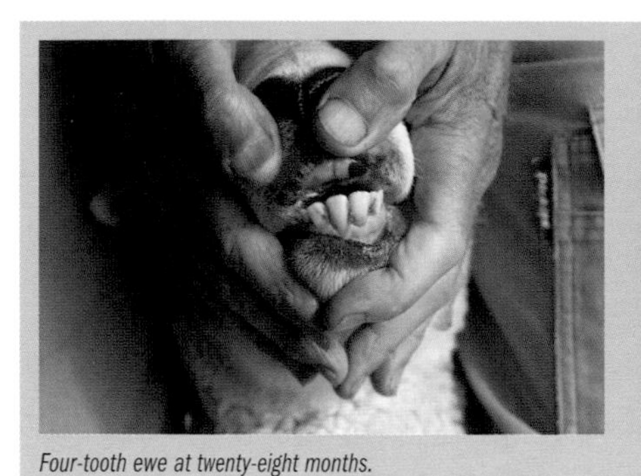
Four-tooth ewe at twenty-eight months.

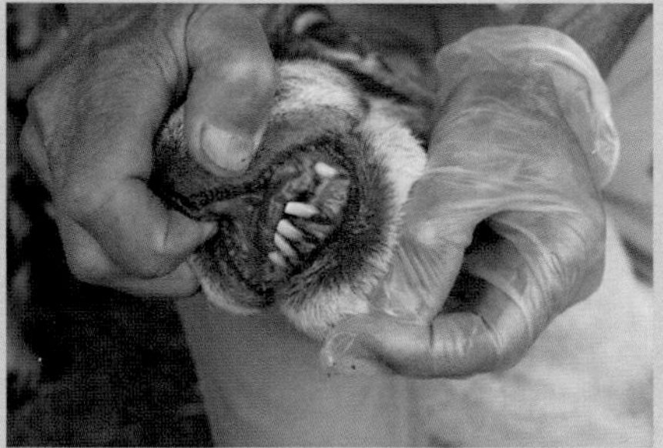
Broken-mouthed ewe at ten years of age.

Condition Scoring Ewes and Rams

Checking sheep body condition scores (BCS) (*see* Chapter 14) is critical for assessing the adequacy of the diet at key stages such as weaning and tupping. Ewes in the correct body condition at mating will have a higher ovulation rate and the potential to rear more lambs. Thin ewes should be moved on to good grass, while fat ewes should be offered poorer grazing. Where grass is scarce, make sure hay is offered ad lib. Thin ewes can be fed up to 0.5kg of ewe nuts per day, but *do not* feed all your ewes 'as a treat' because you will end up with them over-fat and risk having the many serious pregnancy problems that go with that - these include pregnancy toxaemia, prolapses, and over-large lambs that cannot be birthed naturally.

'Flushing' is the practice of moving ewes on to particularly good grazing two to three weeks before introducing the ram, and is to maximise ovulation, improve fertility and increase lamb numbers. If your sheep are in fat, rather than fit condition, this is not advisable.

Ram-to-Ewe Ratio

How many ewes can one ram manage? Well, it depends on how tight you want your lambing pattern to be (how long your lambing will last). A mature ram can cope with thirty to forty ewes, and a ram lamb will manage fifteen to twenty ewes. The issue with a small flock is that the ram will not be satisfied with serving just a handful of sheep, and keeping him contained within fencing that usually works may be a challenge if he can sniff other ewes in the vicinity. With small numbers you may be better off hiring or borrowing a ram just for the tupping season, or using a ram lamb and selling him on once he's done his job.

The Ram

Rams should be in good body condition at the start of tupping; over-fat rams are lazier and have reduced libido and fertility. Check the ram has two testes, that the scrotum is firm and of good size, and that the feet are in good condition, with no lameness. If you are concerned about ram fertility you can have that tested by your vet. If the weather is warm, introduce rams to the ewes in the cooler part of the day. Semen takes six weeks to produce, so rams need to be in good condition at that time, with any stressful treatments such as worming and foot trimming carried out six to eight weeks prior to mating.

You can put a harness on your ram with a coloured crayon that marks the ewe when he mounts her, or use powdered raddle mixed with vegetable oil smeared generously on his brisket between the front legs. Noting the date when he marks ewes is useful for forecasting lambing dates. Changing the raddle colour every fourteen to sixteen days means that you can spot an infertile ram. Rams will only mount ewes that are not pregnant: if he marks one or two with a different colour that is acceptable, but if he marks them all, he is infertile and needs to be replaced if you are to have lambs in the coming season. Once rams reach six years of age their fertility and libido decline significantly.

Using a Teaser

A teaser is a vasectomised ram, not a castrated wether: vasectomy seals the semen tubes so that

Herdwick ram with raddle.

the teaser has testes but is infertile. Putting him with ewes fourteen days before the fertile ram helps bring them into oestrus and reduces the lambing period. This may be useful for a smallholder employed off the holding who needs to tighten their lambing season. The suggested timetable is as follows:

Day 0 – teaser goes in
Day 14 – replace teaser with breeding ram
Day 25 – remove ram from ewes

You can buy a teaser, or have a ram of your own vasectomised by the vet. Choose one with good teeth, body condition and feet, and preferably a shearling so that you can use him for several seasons, and as a ram companion when he is not in work. Rams of a distinctive colour or different breed are often chosen as teasers to avoid any confusion. For best results, keep all rams away from the ewes for a month before teasing/tupping (out of sight, sound and smell).

Tupping

Sheep are mostly short-day breeders, meaning they come into heat when the daylight hours get shorter, from the middle of autumn. There are also long-season breeds that are fertile much of the year and can theoretically lamb twice a year; more commonly they are managed to achieve three lambings every two years. These include Dorset Horn, Poll Dorset, Valais Blacknose and Merino. Charolais ewes come into heat early, around July, and Mules and Suffolks cycle from August or September. The majority of sheep breeds don't usually get going until September, and that can be even later for hill breeds. Our Whiteface Dartmoors and Welsh Mountain Badger Faces come into season in October, but the Herdwicks wait until the beginning of November. We only put ewes to the ram in their second year, as adults.

Within these limitations the timing of tupping depends on when you want to lamb, and for the smallholder who is not focused on the commercial niceties of having fat lambs ready for the butcher by late spring, lambing can be helpfully timed to coincide with the growth of new grass during the spring and when the worst of the winter weather should be over. We put the rams in with the ewes in mid-October to lamb from mid-March. Oestrus cycles in sheep range between fourteen and twenty days, with seventeen as the average, with the ewes on heat for about thirty hours. Rams are fertile year-round.

Removing the Ram

After thirty-four days the ram should have covered two seventeen-day cycles of ewes coming into season, and should be removed to avoid an extended lambing period. If you keep just one ram, a wether (castrated ram lamb) makes a good ram companion if you don't have a teaser.

Some smallholders struggle to find room to separate their ram from the ewes. If they are very docile, you could keep them with the ewes for longer, but once you start to feed concentrates to any ewes expecting multiple lambs in the lead-up to lambing it's best to move the ram away from them or it will steal food intended for the ewes, and can butt them forcibly off the feed, which can risk damage to the unborn lambs.

Depending on the weather and the quality of your grazing, this is the time to start feeding good quality forage to your flock – hay is the best choice. If you are a first-timer, book yourself on a lambing course – lots of large animal vet practices offer these.

Tupping (Torddu Welsh Mountain Badger Face).

MANAGEMENT OF IN-LAMB EWES

Scanning and Separating

We have found pregnancy scanning an essential tool in managing sheep effectively, firstly for appropriate pre-lambing feeding, and then at lambing time we know exactly how many lambs each ewe is having, which can save lamb (and ewe) lives. A skilled technician can tell whether a ewe is pregnant or not thirty days after mating, and from 80 to 100 days individual numbers of foetuses can be identified. Ewes are scanned to check for singles, twins or triplets, and to find out if any ewe is barren. After 100 days it becomes difficult to tell the number of lambs. Gestation is 147–153 days (twenty-one weeks).

Make arrangements after scanning to separate the singles, who are unlikely to need any extra feed other than good hay, unless they are in thin condition. In the eight weeks leading up to lambing, those having twins and triplets will need to be fed ewe nuts (hard feed), either on a rising plane or on a steady plane regime. Any ewe expecting a single lamb but with a low body-condition score should be

Scanning the ewes.

put with the ewes having multiples to feed her up and improve her condition score.

Feeding In-Lamb Ewes

A ewe's feed requirement in spring through to autumn should generally be achieved through grazing, eating around 2 per cent of her bodyweight daily in dry matter; a 60kg ewe needs 1.2kg dry matter daily, equating to 3–5kg of grass, depending on the time of year. We give the ewes expecting a single lamb good hay throughout the winter and their pregnancy, but no hard feed at all to avoid the lamb getting too large, which could result in a difficult lambing. Ewes expecting multiple lambs must also have access to good hay at all times. The doubles and trebles (plus the occasional too-thin ewe expecting a single) are given ewe nuts on a flat plane of nutrition from eight weeks before their due lambing date, of 0.5kg per ewe per day (our sheep breeds are all medium-sized).

Assuming good body condition, small breeds such as Hebrideans, Castlemilk Moorits and Shetlands should be fed according to their size, so 0.25–0.4kg will be plenty. For the heaviest breeds, such as the Suffolk, Teeswater, Cotswold and Texel, 0.5–1kg a day is more appropriate. Half a kilo of hard feed is the maximum that should be fed per ewe in a single feed, so if you have large ewes, split their rations into two even feeds daily. 70 per cent of a lamb's birth-weight is put on in the last six weeks of pregnancy, so the nutritional demands made on the ewe at this point are considerable.

Provide all pregnant ewes with high energy mineral buckets from tupping onwards so they can help themselves as needed. We avoid the 'foot and fertility' mineral bucket as we have found these to cause seriously overgrown hooves.

We have never needed to give hard feed to any of our rams, simply offering them ad lib hay from November until the spring flush of grass.

Out at Grass

Sheep spend most (if not all) of their lives out at grass. Once a sward has been eaten to below 4cm the sheep should be moved on to fresh, longer pasture (ideally 4–8cm). In good growing conditions, if all your grass is nibbled to the nub, it's time to reassess your stocking rates.

Feeding ewes expecting twins.

Feeding Hay

We give hay to the sheep from November until the end of April (depending on the quality of the grass), getting through one small bale per sheep for each month that the grass isn't at its best.

Hay racks in the field should be moved frequently to fresh ground to avoid poaching and over-mucking of the areas around the feeder. This is not for aesthetic purposes but to minimise foot problems and other diseases, such as coccidiosis. Homemade hayracks can be created from pallets, IBC cages and other materials – just be sure they are mobile and keep the hay dry.

Lambs playing around a hay rack.

Round bale feeding in the lambing shed.

Feeding Concentrates

Any change in feed type or quantity should be gradual, and feeding should be at the same time each day. Ensure that all ewes have adequate access to feed to reduce the risk of some ewes overeating and others not having sufficient food. For hard feed, allow eight to ten ewes per 2.75m (9ft) feed trough in the run-up to lambing to ensure that every ewe can eat simultaneously and without being bullied by others; provide additional troughs if any sheep is running from one trough to another without feeding successfully.

If the ground is dry and not mucky you can scatter feed directly on the ground; the larger ewe rolls are easier for the ewes to find. Note that management of a hay rack is not the same as that of feed troughs: hay racks should be kept filled so that sheep can help themselves ad lib, so one hay rack should be enough for a small flock.

Water

All livestock needs access to clean drinking water at all times; it's a complete myth that sheep survive by licking the morning dew. Sheep consume anywhere from 2.5 to 18ltr of water per day, depending on the content of water in their feed and the environmental conditions; their requirements increase greatly during late gestation and lactation. Make sure water troughs are checked and cleaned out when necessary, and are filling as they should.

PRE-LAMBING PREPARATIONS

Vaccinate the flock against clostridial (soil-borne) diseases (*see* Chapter 13) four to six weeks before lambing. Prepare pens for lambing, and sort out your lambing kit. We have all our ewes crutched (the fleece round the tail, bottom and udder is removed)

Crutched ewe with developing udder.

a month before lambing starts, so that we have a clear view of udder development and in due course lambing progress: this has the real benefit that lambs find the teat quickly and easily after birth, and suckle on the teat rather than on clumps of dirty wool.

Lambing (Kidding and Calving) Kit

- Hurdles for creating bonding pens indoors.
- Straw for bedding.
- Hydrated lime or dairy-cow cubicle bedding powder to disinfect bonding pens between uses.
- Water buckets for each pen (calf buckets are perfect), plus a water supply. Have a bucket of water and paper towels available to wash your hands.
- Feed buckets for each pen.
- Saddle hay racks for bonding pens. Avoid hay nets as kids/lambs get tangled up in them and strangle themselves in loose loops.
- Marker spray to identify which lamb/kid belongs to which dam. It's helpful to use a different colour depending on the number of young born alive, so that you know at a glance that any with a red number for example should have two lambs/kids.
- Notepad and pen or whiteboard for keeping records, and which animals need extra care and of what kind – this is critical when you are doing shifts and need to pass on the information to the next person. I use a clipboard with a printout updated daily.
- Heat lamp for poorly newborns (hot water bottles are also very handy).
- Colostrum. This can be colostrum you've frozen, or a powdered type.
- Milk replacer, bottles and teats for feeding milk to any orphans. Also jugs and measuring/weighing kit for preparing colostrum and milk.
- Stomach tubes and syringe.
- Obstetric lubricant and disposable vinyl or latex gloves (short gloves and long-arm versions).
- Carrying buckets for all your equipment.
- Elastrator and rings for castrating ram lambs/bucklings/bull calves (also a small sharp knife for removing rings if necessary).
- Rope halter or head collar and lead if you need to restrain a cow/doe/ewe.
- Lambing ropes.
- Prolapse harness and spoons.
- Medication: pen and strep antibiotic, propylene glycol (in the case of twin lamb/kid disease), injectable calcium/magnesium (in case of hypocalcaemia), oxytocin (to instigate strong contractions in a lazy labour where the cervix is already fully dilated, and to encourage the let-down of milk).
- Iodine liquid/spray or copper sulphate for navels.
- Syringes and needles.
- Torches/head torch.
- CCTV or a livestock monitoring camera means you can observe activity from the warmth of your bed, ready to leap into action if necessary.
- Muck fork and wheelbarrow to remove mucky straw.
- Hay and feed.
- Surgical handscrub/hibiscrub – for assisted birth.
- Digital thermometer.
- Vet's phone number at hand for emergencies.
- Empty plastic feed sacks for removing afterbirth and general rubbish.
- Penknife – for cutting baler twine to opening packets of this and that.

For calving, much of the above is not needed, but do have calving ropes and jack on hand, and iodine or copper sulphate to dress the navel. For farrowing pigs we rarely need any kit.

Once lambing starts keep a careful watch throughout the day and night. We put each mum and their newborns into a family bonding pen for forty-eight hours. Identify each lamb a day or two after birth, so that it can be matched with its mother, who is also marked (spray the same number on the ewe and its lambs, and use a different colour for singles/doubles so you can see at a glance out in the field if a ewe is missing a lamb). Put iodine (we prefer copper sulphate) on the navels of newborn

Ewe and lamb in bonding pen.

However, the longer you have sheep and the more lambings, kiddings and calvings you do, the more you realise the infinite variety of birthing signs. There may be no lying down, no visible water bag, a ewe stuffing breakfast down with gusto with a lamb half hanging out, insignificant pushing (usually indicating a lamb coming breech or posterior; when there is no head to engage in the cervix the pushing tends to be much more half-hearted), and all sorts of other behaviours to keep you guessing.

I wear a watch at lambing if at no other time, to take note of how long certain things are taking: if a sheep is pushing hard for half an hour that's fine, but two hours is too long not to make an internal check. If a sheep has had a water bag out for an hour and is not making any progress, put on a glove and feel

lambs to prevent the ingress of bacteria and resulting disease.

If they are healthy and the weather is good, turn mother and lambs out after forty-eight hours. Check the ewes and lambs several times a day to pick out and treat any sickly or weak lambs, and to make sure no lambs have been separated from their mothers. If necessary, any castration or tail docking of lambs must take place in their first week of life.

LAMBING

If you read textbooks on lambing (please do, they are fabulously informative), they describe the whole process: udder now big and taut, vulva pink and swollen, the previously absurdly wide ewe suddenly less so as her 'saddlebags' disappear as the lambs move lower into the body causing her belly to drop, the ewe getting restless, bleating, pawing at the ground, ears back, probably refusing food, finding a corner on her own, water bag appearing, lying down, up again, down again (ad nauseam), contractions, bag with lamb popping out and whoosh, out it comes, with variations on the theme (*see* Abnormal Presentations below).

Lambing Sequence

The ewe will become restless and the cervix start to dilate. She will focus on a place to give birth (although if in a communal area, this may change). This phase can last between 4–8 hours in experienced dams, and 6–12 hours for first-timers. You may not notice early signs, which makes timings somewhat approximate. You will know for certain once she starts having contractions.	 *Ewe having contractions.*
After plenty of contractions/straining the water bag will emerge, and the ewe is likely to get up and down between contractions. The water bag will break, and the front feet and nose of the lamb will start to show inside a membrane bag. Some ewes do all this standing up, most prefer to be lying down. The feet and nose can appear to pop in and out before making enough progress to be fully delivered – don't worry. In due course the lamb is born. This stage of contractions to birth can take anything from one to four hours, and longer for cows. The waterbag may appear anything up to an hour before the lamb, but can emerge simultaneously.	 *Waterbag visible.*

Waterbag out.

Feet just emerging.

The ewe will lick and clean each lamb as it arrives, and the lambs will find their feet with various wobbly starts and falls, and look for the teat to suckle.

Licking the newborn lamb.

With multiple births (twins, triplets, quads), one lamb may follow the next very quickly, or there may be up to an hour between births.

Newborn twins suckling.

what's going on. Having a feel and noting that all is present and correct means you can probably leave her longer, but if things are jumbled up, tight, or lamb elbows or hocks are locked against the ewe stopping egress, you'll need to take action. (The descriptions that follow for lambing presentations are just the same for goat kids and calves.)

It's always difficult giving timings as you may miss early signs, and there are always exceptions (things can happen much faster or more slowly). Therefore please take the timings as a general guide only. Those keepers new to lambing/kidding/calving are far more likely to want to rush in too quickly and unnecessarily, so sit on your hands and give the ewe/doe/cow time to give birth naturally; busy yourself instead by taking notes of what's happening and at what time.

The placenta will be delivered in due course; this may take several hours. Do not pull on it as this can cause haemorrhaging. I tend to remove the placenta once it is passed to keep the area clean; the dam will often eat it, which is fine, but there is a small risk of choking. If the placenta is hanging from the ewe after a couple of days, do let the vet know – they can remove it manually if necessary.

Shepherds hope to see the 'diving position' of the front feet with the nose not far behind, often with the tongue sticking out too, but there are alternative presentations, some of which the inexperienced can deal with, and others that require support from the vet or an experienced shepherd. Being able to identify front and back legs is an important part of knowing what you're dealing with. The ankle and knee bend the same way for the fore legs, but not for the hind legs.

Always expect multiple lambs – and allow for this possibility even in ewes scanned with a single – and check whether the legs you see or feel are from the same lamb, or not, by running your hand up one leg and down the other to see if they are connected. If using a lambing rope for the head, place the loop behind the ears and through the mouth, not round the neck or you will strangle the lamb.

Abnormal Presentations

There is a range of less helpful presentations – otherwise known as dystocia, or difficult births – some of which are easy to diagnose (a wriggling tail coming first, or the head only, for example), whereas others require an internal examination with clean examination gloves and plenty of lube (obstetric lubrication gel) to assess what position the lamb is in so that you can aid as required. On the subject of lube, it is probably the single most important item in your lambing kit, as it lubricates and eases the path for your hand and the lamb. 'An ounce of lubrication is worth a ton of pressure' is the truism to live by: using plenty of lubrication if you need to assist can make the difference to being able to lamb successfully or not.

Make sure your fingernails are short and that you have removed your rings and watch before you enter a ewe. Always wear a new, clean, disposable arm-length examination glove. If you find these rather large (they are designed for giants with bananas for fingers), put a short wrist-length close-fitting disposable glove on top.

First-time mothers can be more susceptible to problems than mature animals that have given birth previously, so will need greater vigilance; they might be tight, so a squeeze of lube inserted inside the vagina for a ewe struggling to give birth where presentation is otherwise all correct can be very helpful. Obesity and lack of exercise during late pregnancy increase the chances of birthing difficulties.

If you have done as much as you can and are unable to aid the ewe, call your vet or an experienced shepherd; under no circumstances should you get a host of inexperienced people to 'have a go'. A vet can help sort problems if given time, so don't leave it too late to make contact, and they alone can perform a caesarean if it's needed. As a first-time lamber don't be nervous about calling your vet just because you think you should be able to cope: use their skills and learn from the experience so that you feel better able to manage on your own in the future.

Head and One Leg Only (Front Leg Back)

If the lamb is small you can birth this presentation without correcting the position of the backwards leg by bringing forwards the head and protruding leg. If the backward leg is crooked and locked against the ewe's pelvis you will need to cup your hand round the lamb's hoof to avoid damaging the ewe, straighten the leg and bring it forwards. If the lamb is large you will also need to bring the hidden leg forwards before you can birth it. If the head is not too far out push it back first to give you more room to move the leg in the uterus rather than in the confines of the birth canal.

Head Out Only (Both Front Legs Back)

With both legs back/head out only, there is a real danger that the lamb's head will swell as the ewe contracts, which can be fatal; do not ever try to pull the lamb out by its head alone. If there is room, reach in and draw out a leg. You may be able to birth it as for head and one leg only, or reach in again and draw out the second leg and birth the lamb. If there is no room to bring out the leg(s), attach a lambing rope around the lamb's head behind the ears and through the mouth, and with the help of your trusty lube push the head back in. Find and bring out the legs and then the head using the lambing rope, and deliver the lamb.

Elbow Lock

With an elbow lock, put your hand in and hook out the leg(s) one at a time and deliver the lamb.

Head Back

If the head is back you will need to bring the head forwards in order to deliver the lamb. Use a lambing aid/rope to stop the head flopping back into its original position.

Breech and Posterior (Coming Backwards)

With a breech (bottom only) or posterior (hind legs only) presentation, speed is needed once your pulling has broken the umbilical cord, so that it is out of the mother and all fluid is cleared from the mouth before it takes its first breath. Do not attempt to turn the lamb into the normal presentation. With a breech you will have to go into the ewe to bring out the back legs and then deliver the lamb. With posterior, check the legs are from the same lamb, straighten the legs and birth it quickly, pulling downwards. If the lamb gets stuck, lift one leg upwards at a time to unlock it. Take particular care in both cases to swiftly clear any mucus from the lamb's mouth and nose; use your fingers to remove gunk, and hold it upside down by the back legs so that gravity further clears any remaining mucus.

More than One Lamb Coming at Once

Lambs can come together, both facing forwards, both facing backwards, or one forwards and one back. You will need to investigate the situation so that you know which head and legs belong to which lamb, then push one lamb back into the ewe, and deliver first one and then the other.

Mismothering

When lambing in a communal space ewes can be very active in mothering a newborn that is not their own, particularly when the dam is focusing on delivering a second lamb. Once the first lamb is born and licked by its mother so that she knows it, carry the lamb in front of her nose – keep it low and in sight, make bleating sounds, and if the ewe moves away, put the lamb down and retreat until she comes to her lamb again; like this lead the pair into a bonding pen. This will shut out the hopeful mismothering auntie and allow the family to bond in peace.

This is not a case of misplaced anthropomorphism: if the lamb is stolen by another ewe she may abandon her own lamb when it is born, or the one that she has mismothered; the lambs may not receive their colostrum; it can leave you with lambs that need to be bottle fed; and you can end up with

Normal presentation.

Head and one leg.

Head only.

Elbow lock.

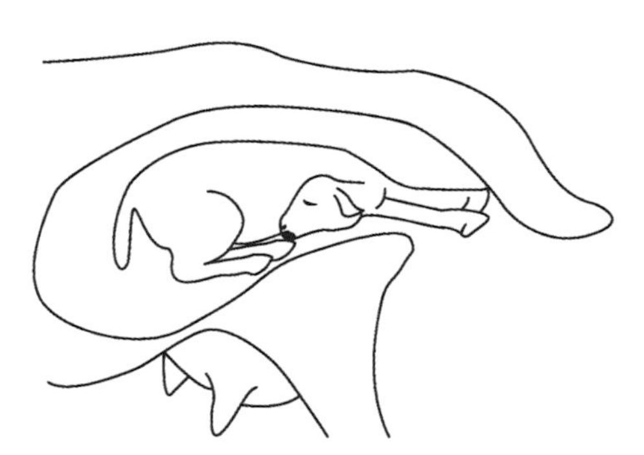

Head back.

Breech.

Posterior.

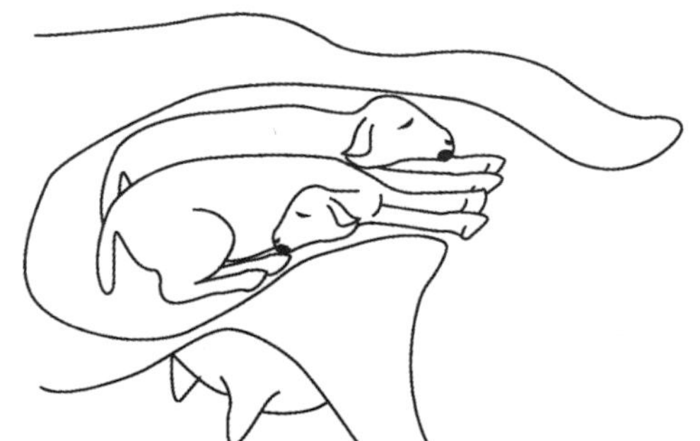

Two coming at once, one head back,

One coming forward, one backward.

too many lambs on one mother and not enough with another.

Care of the Newborn Lamb

The majority of lambs are born without any assistance, and within minutes are up on their feet instinctively seeking the ewe's teat. Some are a little slower, and this can depend on breed, as well as how easy their birth was. Check that the lamb has fed from its mother within the first two hours. If you see a lamb asleep in a pen don't just assume that it's fine: give it a gentle poke – it should stand up and stretch if all is well.

Within an hour of birth, carry out a quick examination of the lamb:

- Check for swellings on the lamb's body, head or limbs, especially around the umbilical cord, which may indicate a hernia.
- Check the anus to make sure there is an opening and that it's not covered by a thin membrane, which will prevent it passing faeces. If imperforate this needs to be opened by the vet.
- Check the lamb is able to stand: lack of coordination may be a symptom of underlying disease or infection.
- Tucked-under feet or contracted tendons often improve with no treatment; if more severe, make a splint from lolly sticks, cotton wool and vet wrap to help straighten the legs.
- Check the lamb's mouth and lower jaw – if severely over- or undershot it will have difficulty suckling and need feeding by stomach tube.
- Examine the lamb's eyes for entropion: this is where the eyelid is turned in, causing the lashes to rub the eye, which can ulcerate. This can often be cured by pinching and rolling out the in-turned eyelid. If not, it will need 1ml of penicillin injected horizontally into the eyelid – not a job for the inexperienced, but a simple job for the vet, or the keeper once they are shown how.
- Within the first few hours the lamb passes sticky black faeces (meconium). Once meconium is passed, the faeces will be orange and can be very sticky – keep checking that this fudge-like poo doesn't plug up the lamb's anus, and if it does, wash it carefully away with warm water. At around ten days old it will start passing miniature sheep droppings.

Lambing Pens

In the lambing pen check the ewe's udder for milk supply and potential problems, such as mastitis (a hard, hot udder from which it is difficult or impossible to strip out milk). Each teat can be stripped to remove the wax plug, and lambs should be monitored to make sure they nurse. Lambs that have nursed will have a full stomach, which you can feel by palpating. Lambs that have not nursed should be assisted.

The size of the lambing pen can be varied depending on the size of the ewe: 1.2 × 1.8m is large enough for the shepherd to get in with the sheep to carry out any necessary tasks. Pens should be clean, dry and well bedded up, and cleaned between ewes. Having one pen per five ewes may be adequate for small flocks; make sure you have enough pens to keep new sheep families together for forty-eight hours, with space for more as the ewes give birth. Having a few more hurdles than you think you will need is always advised for pens, and to separate off areas if necessary. Part-meshed lambing hurdles are particularly useful in stopping lambs escaping.

Heat lamps should be used for hypothermic lambs with caution; to avoid fires, hang these in the corner of the pen a metre above the bedding, and protect the cable from chewing.

Tubing

In some cases a lamb is unable to nurse even with assistance, if it is small, weak, chilled, rejected by the ewe, or injured. In this case stomach-tube feeding is necessary to get colostrum into the lamb. To do this, sit the lamb in your lap and with its head in a natural position, insert the tube in the side of the lamb's mouth, following the roof of the mouth down into the throat. Don't force the tube down, but allow the lamb to swallow as the tube goes down the oesophagus. The tube can be felt on the outside of the neck as it is inserted down into the stomach, and will go in about 30cm. Although it is difficult to get the tube down the trachea (windpipe), check to see if air is being expelled (listen, or moisten the end of the tube to see if bubbles form). After inserting the tube, attach the syringe and fill it with colostrum, allowing the colostrum to go down by gravity; don't use the plunger.

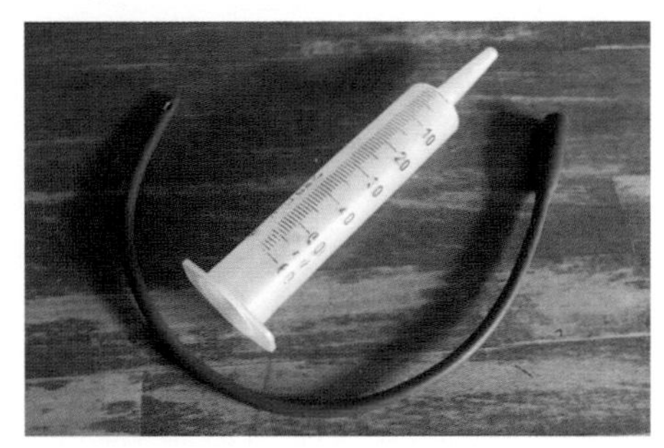

Stomach tube and syringe.

Lambs should receive 10 per cent of their bodyweight in colostrum within twenty-four hours of birth, ideally consuming half of this within four to eight hours. After initial tube feeding, many lambs will respond and begin to nurse on their own. If not, the lamb may need to be tube fed every two to three hours.

Fifteen hours after birth, colostrum production ceases and ewes produce milk. If you have ewes with single lambs, you can harvest colostrum for future emergencies: milk about 100–150ml colostrum from each ewe, put it into a sterile plastic container, and freeze it; it can be gently reheated for use. Never boil or microwave colostrum as this will destroy the antibodies and protein.

Hypothermia (low body temperature) is the greatest cause of death of newborn lambs, so if you see a young lamb curled up in the corner of the pen don't just assume it's sleeping off a feed. Is its abdomen full or empty? Will it raise its head or stand? A hypothermic lamb has a body temperature of 37.5°C or below (normal is 39–40°C). The more quickly a ewe licks her lamb dry, the less vulnerable it will be to chilling. Don't attempt to bottle-feed a cold or weak lamb, as there is a danger of it inhaling liquid into its lungs, which will drown it. A lamb less than six hours old should be warmed, and once it is able to hold its head up, should be stomach tubed with colostrum.

Adoption

The optimum result is to have two lambs for each sheep (if she has enough milk). With triplets, unless the dam is very milky, you should adopt one of the triplets on to a ewe having a single or whose lamb has died; if a ewe dies, you will want to adopt its lamb on to another ewe. The aim is to minimise the number of bottle-fed lambs, as these rarely do as well as those feeding from a ewe. There are various methods for doing this, described below.

Wet Adoption

Fill a clean shallow bucket with 10cm of warm water and 100g of salt; mix together. Take your adoptee lamb and tie its two front legs and two back legs together (not too tight), then place it in the bucket. Take the bucket/lamb to the lambing ewe, and lamb her lamb straight into the bucket on to the adoptee. Cover both lambs in the juices and put the ewe in a pen with both lambs out of the bucket. Leave the orphan lamb's legs tied so it simulates the behaviour of a newborn and allows the other to get the good colostrum. The salt encourages the ewe to lick and so mother the adoptee, and makes both lambs smell and taste the same. Once the lambs have been licked dry and the newborn has had its fill of colostrum, untie the orphan lamb's legs and observe that it is accepted.

If time is short and there is no time for warm water and buckets, lamb the ewe having a single on to a clean, empty feed sack, tie the legs of the adoptee and rub it in the birth juices, and continue as above.

Adopted lamb in the dead lamb's skin.

Dry Adoption – a Jacket for Your Lamb

Although not for the fainthearted, this process rarely fails. If a lamb dies and is no longer wet from birth – so not stillborn, but hours or a few days old – skin it and put its jacket on a surplus lamb (for example, the largest of triplets) and give it to the dam whose lamb has died. Make sure the lamb's rear end is well covered by the skin, as that is the first area the ewe will sniff and lick. Once successfully adopted, remove the skin after a couple of days – and definitely don't leave it on once it has started to rot/stink. If a lamb is stillborn and is wet, you can use the wet adoption method above by rubbing the adoptee on the dead lamb.

Wet adoption.

Adopter Crate

An adopter pen that secures the ewe's head and stops her butting an adopted lamb away is a useful tool, and safer than using a halter that might strangle her. This allows the lamb to suckle the ewe even if she's not keen to adopt at first. For us it's the least preferred approach – we prefer to skin a dead lamb, considering this to be a more humane option for the sheep.

Adopter pen.

Supplementary Feeding of Lambs

Lambs should only need their mother's milk and grass/hay, unless they have special needs (if they are orphaned or poorly), in which case they may need bottle feeding and supplementary creep feed. We avoid giving creep feed ad lib, preferring to give small amounts to avoid the lambs gorging themselves, which can all too easily result in bloat, which is frequently fatal. If you have any orphans or bottle-fed lambs, follow the advice on the milk replacement you have bought. Lambs should have been eating creep feed (approximately 0.25kg daily) for at least ten days before weaning from milk replacer, and be at least 2.5 times their birth weight. By the time they are two months old and if the grass is good, you can stop feeding orphan lambs hard feed.

We don't give any hard feed (such as pelleted lamb creep feed or barley) to lambs that are suckling well off their mothers, so they develop on their mother's milk and finish slowly on grass – this improves the taste of the lamb and lays down meat rather than fat.

Ear Tagging

As long as the legal requirements are met, the timing for tagging lambs varies depending on preference. If you have pedigree stock registered with a breed society it's important that the dam of the specific lamb(s) is known. We tag when the lambs are forty-eight hours old as part of our post-lambing turn-out, making a note of the lambs' ear-tag numbers and that of the corresponding dam so parentage is established in our records. If you have properly numbered the ewe and her offspring with spray you can ear tag lambs at the time you vaccinate, when the lambs are about a month old and numbers are still clear on the fleece. Leaving it later can make it difficult to be certain which lamb belongs to which ewe, unless you have a very small flock and know each one and its family members intimately, or have a great eye, as some impressive shepherds do, for recognition.

Avoid tagging in warm, moist conditions when there is a risk of flystrike. Tag lambs on the upper edge of their ear, a third of the ear length from the head, with the smooth part with the male pin on the back of the ear and the knobby, female part inside the ear. Make sure you leave some growing room for the ear, allowing 0.5cm growing room or more for breeds that develop particularly large ears (Suffolk

Creep feeder.

Lambs in homemade creep area.

Whiteface Dartmoor family numbered with marker spray.

and Border Leicester, for example). Lambs have a ridge of cartilage along the top of their ear so tag below this and not into the cartilage itself.

There are many designs of tag available; choose tags that are smooth on the outer side – some manufacturers make ones with knobbly bits on both sides, which simply act as ear-rippers when in contact with fencing, hedges, other sheep and any other thing they might rub up against. No ear tag is entirely infallible against ripping an ear, but you can minimise the likelihood. For the visual tag choose colours where the numbering is legible (dark-coloured tags are particularly challenging to read in anything but bright sunlight). We use a different colour visual tag each year to tell at a glance the age of the sheep. It is particularly helpful when lambing to know from the far side of the lambing shed if you are dealing with a (tight, possibly nervous) first-timer or an old girl who may tire and need a bit of help.

WEANING

At sixteen to twenty weeks (at around 20kg), the lambs will be weaned from their mothers and put on the best pasture to fatten. By the end of August the first of them will be 'fat', weighing 25–40kg (depending on the breed), and ready for the butcher.

At weaning, put the ewes on poor pasture to dry up their milk – a field that has recently had hay cut from it is ideal. Uncastrated ram lambs become sexually mature at five months, so separate them from all females (lambs and ewes); wethers (castrated ram lambs) can be kept with ewe lambs.

SHEARING

If the ewes have been crutched pre-lambing, this keeps their back ends clean. By June the wool on the adults should have risen, lifting and coming away around the neck, and it's time for them to be shorn. The optimum time for shearing depends on breed and your location – a hill breed such as our Badger

Ready for shearing.

Battery shearing kit.

Faces may be two or more weeks later than lowland breeds, and we shear in mid-June. The further north you are, the later will be your shearing season. If you are fit and keen, go on a course to learn how to shear. If you are not, turn the job over to a professional: their skill and speed will massively decrease any stress on the sheep and on you. There are plenty of shearers offering their services to smallholders with small flocks.

If you have more than one ram, even if they've been cohabiting in perfect harmony, once they've been shorn they will no longer recognise each other's smell and can fight. This can be serious, sometimes resulting in fatalities. To manage this, once shorn put the rams together in a small pen of hurdles so they cannot take a run at each other. There will be biffing and bruising, but this will ease over a few days. Enlarge the pen after a day or two to see how they behave, and if they are calm they should be fine to go back in their field. The same process has to be gone through again post-tupping if they have been separated and are being reintroduced.

If lambs have mucky bums, investigate why: (do faecal egg counts – FECs) and treat accordingly, and dag the lambs (clip the dirty wool off their tails/rear ends) to remove any mucky wool around their bottoms to keep the flies away to prevent flystrike (*see* Chapter 13).

In September you might choose to have any lambs shorn. This keeps them safe from being entangled in brambles, and encourages them to eat well, putting on weight over the winter. By the time the weather is truly cold they will have grown a new good covering of fleece.

Shearing.

11 Goats

The goat is an ideal smallholder animal, providing milk (therefore cream, cheese, butter, yogurt, ice cream and soap), meat, skins, fibre, horn and plenty of interaction. They can clear scrubby areas, and don't need vast tracts of land to thrive – although as browsers rather than grazers you will need to provide them with hay and browse, and having a shelter is a necessity.

SUITABLE BREEDS FOR THE SMALLHOLDER

If you're after goat meat, then a Boer or Boer cross is ideal, although all surplus males including those from the various dairy breeds will give you great meat; however, a Pygmy isn't really suitable, as it doesn't yield any more than a few mouthfuls. Your choice of dairy breeds is fairly broad: a couple of Golden Guernsey or Toggenburg does will produce a family's milk requirements, with a little surplus that can be made into cheese. If regular dairying, or having surplus milk to feed growing pigs or calves is what you're after, the British Toggenburg, British Saanen, Saanen or British Alpine have higher milk yields; and for yogurt, butter and cheesemaking the Anglo Nubian, whose milk is high in butterfat, is a great choice.

Purpose-Bred Goat Types

Types	Description	
Meat Boer	Known as the world's best meat breed. Used as a sire for dairy herds to produce kids with a viable meat carcase.	*Boer doe.*
Dairy pure breeds Golden Guernsey Saanen Toggenburg **Native dairy** Anglo-Nubian British Alpine British Guernsey British Toggenburg British Saanen	For milk production and dairy products. *Golden Guernsey:* Small goat with moderate milk yield. *Saanen and Toggenburg:* High-yielding milk goats. *Anglo and British:* Dairy breeds created by crossing British natives with imported goats.	*British Alpine.*

Types	Description	
Native English Old English British Primitive	Dual purpose, suitable for milk and meat. *English:* Utility smallholder breed. *Old English:* Primitive and feral goats found mostly on hilly areas of the UK.	 *English goat.*
Fibre Angora Cashmere	The Angora produces mohair (not to be confused with angora wool, which comes from Angora rabbits). There is no specific cashmere breed; it is the name given to any goat with a soft downy undercoat selectively bred to produce a yield of cashmere.	 *Angora.*
Conservation grazing Bagot	Small, long-haired, hardy, nervous, not keen on being handled. At risk rare breed.	 *Bagot.*

Types	Description	
Non-productive/pets Pygmy	Hardy, genetically small, compact goat originating from Africa.	*Pygmy.*

THE GOAT KEEPER'S YEAR

Although there are similarities with the shepherd's year, a goat's needs are not quite the same as for sheep, and there are fundamental differences, such as does coming into season every twenty-one days (ewes cycle every seventeen days) in the autumn. The timing of most events depends on what part of the country you're in, the breeds you have, and when you kid if you choose to breed. The calendar below is based on February kidding. Year round you should check feet, body condition, worm counts and parasite burdens, and deal with these accordingly.

FEEDING

Goats are perfectly designed to digest fibrous plant material rather than grain, which can create all sorts of digestive, foot and bladder problems. Keep things as natural and as suitable to their digestive system as possible, with concentrated feed supplements (hard feed or compound feed such as coarse mix) made available only at those times when the body has additional demands that need to be supported. This includes:

- when a goat is producing milk for the household
- in the late stages of pregnancy and early weeks of lactation
- for young growing goats, but only if needed
- to any underweight bucks or does as they approach the breeding season.

Good quality hay is essential for goats year-round, and when goats are housed, browse should also be provided.

Hay mangers or racks should be positioned so that goats cannot climb on to them and spoil feed with their feet, pee or poo. Use mangers or racks with smaller mesh so they can't pull too much out at one time, as anything dropped on the floor will be wasted; also smaller mesh minimises the risk

The Goat Keeper's Monthly Management Tasks

Month	Management tasks
January	Scan does 80–100 days from breeding date. Start feeding concentrates to those expecting multiple kids and any thin singles fifty days before kidding if their body condition score warrants it. Check for lice and treat if needed. Vaccinate does four weeks before kidding (include doelings, wethers and bucks too if timing is appropriate).
February	Trim hairy does around the back end. Clean the kidding area and set up bonding pens and kidding kit. Keep a close eye on heavily pregnant does. Take FEC samples from does to see if worming or flukicide is required post kidding. Kidding starts. Ring (castrate) bucklings within seven days, and call a vet to disbud kids if required.
March	Kidding continues.
April	Post-kidding treatments and checks: treat for lice, worms and fluke if required. Give kids preventative treatment for coccidiosis if advised. Trim feet if needed. Keep an eye on does for mastitis. Ear tag kids. Process year-old meat goats via the butcher when they are at their peak. Wean entire bucklings at six weeks.
May	Vaccinate kids at 10–12 weeks (first dose). Keep alert for pink eye and orf outbreaks. Wean wethers and doelings at a preferred time. Sell weaned kids if appropriate. Give kids preventative treatment for coccidiosis if required.
June	Vaccinate kids at 14–16 weeks (second dose). Take FECs for kids, and worm as required. Protect fibre breeds against blowfly strike (this may be necessary earlier or later).
July	Ensure potential breeders have a good body condition score in the lead-up to mating. Make sure doelings cannot be accessed by entire bucks. Give adults their six-monthly vaccine booster. Check condition of buck(s), and fertility test if desired. Give kids preventative treatment for coccidiosis if required.
August	Pre-mating treatments and checks. Take FECs, treat for fluke, worms and external parasites if required. Trim feet if necessary. Teeth, tits and toes check for does and bucks. Body condition scoring. Keep breeders, send off the rest for mutton.
September	Annual CAE blood test before does are bred (also CL and Johne's if desired). Put the buck(s) to the does (for February kidding).
October	Remove bucks after two cycles (42–45 days). Give kids preventative treatment for coccidiosis if required.
November	Start to dry off dairy does.
December	Make a list of all fencing and housing to be fixed (a year-round task). Give kids their six-monthly vaccine booster.

The Differences between Goats and Sheep

	Goats	Sheep
Agility	Off-the-scale agile – supreme athletes, from limbo dancing under narrow gaps to scaling walls.	Agile, in particular hill and mountain breeds.
Vocality	Can be very conversational, or fairly quiet; when a doe is in heat they are very noisy. Anglo Nubians are known for being particularly vocal. Bleating in otherwise quiet goats can indicate a problem. Not feeding treats can help avoid goats screaming every time you pass by.	Mainly quiet unless under stress or calling to young. When a sheep is bleating you investigate in order to resolve a problem (either that or you're late with their feed, which is normally only required around lambing time).
Odour	Bucks in the breeding season stink. A few people don't mind the smell – most find it quite vile.	Rams in the breeding season are fairly pungent, a smell that you can detect as you approach them, but it's not overpowering.
Behaviour	Goats can be very pet-like in their behaviour, keen to be with and follow their keeper. Can be destructive and extremely inquisitive about everything in their environment. Head butting is normal behaviour between goats – but don't allow them to butt you.	Apart from those that are bottle fed as lambs and petted, sheep look to their flock for companionship. Individuals can be friendly, but the natural sheep prey instinct is strong.
Social organisation	The herd has a natural hierarchy and structure – goats are social creatures and shouldn't be kept alone.	The flock has a natural hierarchy and structure – sheep are social creatures and shouldn't be kept alone.
Feeding habits	Browsers rather than grazers, with browse making up approximately 60 per cent of their diet. Access to hay and mineral supplements is required year round. They prefer to eat at, or above head height. Require high levels of copper.	Grazers rather than browsers. Browse is approximately 10–15 per cent of their diet. Hay need only be provided in winter months. They prefer to eat from the ground (although happy to stand on their back legs to reach tasty things in trees – fruits or leaves). Copper is toxic to most sheep breeds.
Cost	Tend to be more expensive to purchase and keep.	Less expensive to purchase and keep (there are exceptions for highly desirable individuals).
Commercial potential	There is a growing and developing market for goats, their meat and their dairy produce. Lack of availability keeps prices high, with 105,000 goats in the UK.	Existing, well developed market. Prices fluctuate nationally and globally. There are 23 million sheep in the UK.
Milk yield	Dairy doe: 3–5ltr per day.	Milking ewe: 2ltr per day.
Gestation period	Approximately 150 days (143–157).	Approximately 147 days (144–152).
Season	Goats come into season every 21 days in the autumn until they are bred.	Sheep come into season every 17 days in the autumn until they are bred.
Foot care	Goats may need their feet trimmed regularly, but only when necessary.	Only trim sheep's feet when genuinely overgrown. Inappropriate trimming increases the risk of bacterial spread and long-term hoof injury.
Prolificacy	Goats regularly have multiple births, with triplets and quadruplets not uncommon.	It has been known for ewes to give birth to up to six lambs, but this is most uncommon. The usual number of lambs is two, with frequent singles and triplets.
Fencing	Excellent fencing is required.	Good stock fencing is required.
Horns	There are no polled (hornless) goat breeds, although a few individuals are born without horns. Do not confuse polled with disbudded: many kids are disbudded in the first week of life (by the vet) so that their horns don't develop. Angoras and Boers are often kept horned, while dairy breeds are mostly disbudded.	Many sheep breeds are naturally polled, in some breeds the males are horned and females polled, and others have horned males and females, and sometimes multiple horns (four or six). In horned breeds, the males have larger, thicker horns than the females.
Fibre	Cashmere and mohair from fibre breeds.	Wool from most breeds, apart from hair and self-shedding sheep.
Meat	Delicious, low in fat.	Delicious. Nearly twice the calories of goat.
Tails	Naturally short-tailed.	Apart from the Northern European short-tailed breeds, sheep have naturally long tails (though these are often docked shortly after birth).
Shelter	Goats have little insulating fat (and unless a fibre breed, thin coats), so must be provided with a built shelter against wind and rain.	In all but the most severe weather sheep manage well with the shelter of trees and hedges because of their good wool and fat insulation.
Water	Goats will not drink water that is in any way contaminated – clean water must be provided at all times.	Sheep prefer (and should be given) clean water but are less fussy than goats if leaves or other natural detritus collects in their drinking trough.
Parasites	Goats do not build resistance to intestinal parasites.	Healthy adults build significant resistance to intestinal worms.

	Goats	Sheep
Medication	Most medications are not licensed for goats so a knowledgeable goat vet is essential for prescribing appropriate type, quantity and frequency of medication. Pour-on treatments are not advisable.	All large animal vets are familiar with sheep, and there are many medications licensed for them.
Lifespan	8-12 years. Bucks 8-10, Does 11-12, and longer if not bred after 10 (up to 16-18 years). The record is 22 years.	10-12 years. The record is an astonishing 28 years.
Space requirements	2-3 goats plus their kids per acre. Lower numbers of 1-2 goats per acre on areas of brush.	3 to 5 ewes plus their lambs per acre.
Entertainment value	If you have a GSOH, high.	They have their moments!

of a goat or kid getting itself stuck. Lidded racks are particularly useful for minimising spillage. Avoid haynets of the type used for horses, as goats can get tangled in them, and kids can strangle themselves.

Heavier goats consume more per day than lighter ones, and does that are lactating consume more feed than those that are not. Manage feed according to the individual body condition (*see* Chapter 14), as it is as problematic to have thin goats as it is to have them too fat. Goats are surprisingly fussy eaters, so make sure they find their feed palatable, and that it is fresh and clean. Forage and browse, fresh or preserved, should form 60–100 per cent of a goat's diet.

Hay rack with step access.

Goats will graze grass as well as browse on hedges, shrubs and trees, and taller pastures with a sward height of around 7cm (3in) and more will help minimise worm burdens by keeping them feeding above ground level. Preserved forage will be mainly hay made from grass, but can include tree hay or dried nettles. Goats will need 1–2kg of hay daily and more if other feedstuffs are not available, with milkers needing nearer to 3.5kg of hay per day.

Browse (leaves, soft shoots, bark and fruits of woody plants from trees and shrubs) from ash, elder, elm, hazel, hawthorn and willow are all safe choices and much enjoyed by goats, and as willow grows by the simple expedient of putting a piece in damp ground, you could profitably create a willow bed to cut from for your appreciative goats. Quality straw used for bedding will often be nibbled when freshly scattered, but it can also be put in mangers as part of the forage feed to bulk out the diet for those who need to avoid putting on more condition.

Feeding the Pregnant Doe

Females when they are mating need to be in good condition, and this should be maintained throughout pregnancy. Give her the best quality hay, particularly in the last third of her pregnancy in the lead-up to kidding, and ensure constant access to minerals (calcium requirements increase significantly in late pregnancy) and clean water. To meet her energy needs and maintain her condition it is probable that some concentrates will need to be fed in the last four to six weeks of pregnancy – but ensure she does not get over fat.

Feeding the Dairy Doe

Dairy does need to eat around 5 per cent of their bodyweight daily in dry matter to maintain good milk production. A lactating doe that weighs 70kg and produces about 4.5ltr of milk will need to eat around 3.5kg of dry matter a day, of which the roughage-to-concentrate ratio should be 50:50. Does of the same weight producing nearer to 9ltr per day will need to eat 6–7 per cent of their bodyweight in feed (4–5kg). Dairy goat pellets are available, but consider feeding dairy cattle nuts instead: cattle feed is not only cheaper, it has a better mineral content that suits goats well. Dairy nuts are given as 0.25kg feeds as the doe is being milked (totalling 0.5kg across their two milkings), with a further 1kg fed back in the barn alongside their hay. A milking goat should be fed concentrates from about four to six weeks before kidding, building up to at least 0.5kg at kidding.

Feeding Males

Males, both entire and wethers, should not be given any concentrated feed or grain in order to avoid urinary calculi, a build-up of calcium crystals or stones causing blockages in the urinary system, which can be fatal. Only if a buck is underweight on the lead-up to mating should he be given concentrates: 0.5–0.8kg of a 14 per cent protein feed in the two to six weeks prior to mating.

Supplements

Goats have a high mineral requirement, and mineral supplements should be offered ad lib. These will include calcium, cobalt, copper, iron, magnesium, phosphorus, selenium, sodium, zinc and others. Goats can be copper deficient so this is particularly critical. If you have sheep, ensure they cannot access goat mineral supplements or feedstuffs, as copper is toxic to them. Added minerals will not be needed for those goats eating dairy nuts.

Goats should have access to a salt lick or lump rock salt at all times, and they can be given a small daily amount of seaweed meal as it is full of minerals including iodine and phosphorus, which is critical for the in-kid/freshly kidded goat.

KEEP CONCENTRATED FEEDSTUFFS AWAY FROM GOATS

You may not be feeding concentrates to your goats, but may still have feed on the smallholding for other livestock. Keep feedstuffs in a place that the goats cannot access, and in containers that they cannot breach: the consequences of goats eating large quantities of hard feed can be, and often are fatal through rumen acidosis (grain overload). Any spilled feed should be cleared away before goats can eat it.

Water

Fresh, clean water must be available to all goats at all times, so use drinkers that minimise the risk of contamination: keep drinkers small where practicable, and put them at a height or position where all goats can drink easily, and which frustrates their desire to drop their poo in it; and have enough drinkers so that a bullying goat can't stand guard over a single drinker. Goats can drink anything from 2 to 18ltr of water per day, and the average for a lactating goat is 4–8ltr (for every litre of milk they produce they will need 1.5–2.2ltr of water).

THE DIFFERENCES BETWEEN GOATS AND SHEEP

If you are pondering whether to keep sheep or goats, the table in this chapter (pages 109–110) might help you think things through. You can always keep both, of course, but they should be kept apart and on different ground because of parasite issues, mineral differences and fencing requirements.

Goat water buckets.

Goat water trough.

Bowl drinker with step.

HOUSING

There are many choices when it comes to providing shelter for goats, depending on preference and budget and how you see your goat-keeping life. A simple three-sided open barn with permanent access to a field might suit your purposes, large enough to house an undercover hay rack and be closed off with hurdles when you wish to contain the goats for kidding or handling. You may have existing stables, a garage or cattle shed, or some other building that you intend to convert for goats, or maybe you are planning to build something new for your lucky herd. For the most part, goats can be turned out during the day unless the weather is very bad.

Whether a conversion, a make-do-and-mend or a purpose-built arrangement, bigger (and higher) is always better, and there are some important considerations to take into account, as described below.

Space: Indoor space allowances depend on whether goats are individually penned or live in a communal barn area. For the former, allow 4sq m of floor space per goat, and for communal housing a minimum of 2sq m per goat.

Goat shed.

Two-tier goat barn.

Horned goats: In communal housing any horned goats should be housed separately from those without horns.

Ventilation: Your building must be well ventilated to ensure a flow of fresh air and the natural removal of stale and moist air while being free of draughts. Position doors/openings away from prevailing winds, and install a vented ridge at the apex of the roof.

Materials: The materials used need to avoid moisture condensation on roofs, walls and other surfaces, as moist air brings a high risk of pneumonia. Make sure housing is free of anything on which a goat could hurt itself, including loops of baler twine, protruding nails, and so on.

Size does count: Larger, taller buildings offer better air space, which supports good respiratory health. The absolute minimum height has to allow good standing access for cleaning and tending to the goats, and for them to stand upright on their hind legs at full stretch. If you can, allow a minimum 2–3m at the ridge, and taller is better.

Floor surfaces: Preferences for floor surfaces vary. Earth floors allow urine to drain and can be kept hygienic with a regular covering of hydrated lime. Concrete floors are easier to muck out, are less attractive to rats, can be made slightly ridged to avoid slipping, and can be installed with a slight slope to allow draining. Straw or shavings for bedding will be needed for loafing areas whatever the floor surface.

Dividing the space: The design of internal partitions and gates will vary depending on how you wish to split things up, but a minimum height of 1.3m is recommended. Meshed gates rather than solid partitions allow individually penned goats to see each other. If you keep a buck you will need separate housing for him.

Raised beds: Because goats are so agile and draughts are often at ground level, raised bed areas are ideal for loafing and sleeping.

Play area: For kids and young goats a play area is much appreciated; this might be a sturdy table they can lie on or under, a tree trunk, or something more imaginative with steps, ramps and different heights.

Electrics: Electrical wiring, switches, lights and sockets must all be kept well out of reach of the agile goat. Protect wiring with conduit, and use metal-clad sockets that are waterproof and rated for outdoor use.

Milking parlour: You can create a discrete parlour space in a barn positioned and designed so that it can be kept really clean, or make a small milking parlour used solely for the purpose. The floor of the area you milk in should be very easy to clean, and must survive being hosed down after every milking; concrete is the simplest option. Milk residue on the floor attracts flies, which is unpleasant for the goat and the milker so needs to be removed. The walls and ceilings should also be cleanable, and kept free of dust and cobwebs.

Raised bed.

Play area.

BREEDING

Not all goats capable of breeding should be bred; if you have goats with poor conformation or temperament, or inherent health issues, don't breed from them. If you don't have an entire buck of your own you may be able to run your does with a neighbour's buck, or you might have to travel some distance to source a suitable male. Owners of stud goats will expect that any does brought to be mated will have a certificate to confirm they have tested negative for CAE (caprine arthritis encephalitis).

Boer buck.

At a year old a buck should be able to manage up to ten does over the breeding season; by the time he is two this increases to around twenty-five, and a mature three-year-old copes with up to fifty does. During the mating season a buck loses much of his interest in food, so needs to be in very good body condition beforehand. Overfed, fat bucks are lazier and tend to have reduced libido and fertility. Check that the buck's scrotum is firm and of good size, that there are no foot problems, and that he is fit.

Doelings come into season at around six months of age, but it is advisable not to breed them in their first year. Wait until they are more mature and coming into their second autumn at around eighteen months old, when they are better grown. If you don't want young or other does to be bred they must be kept at a distance from the buck; there are many examples of bucks and does breeding successfully through a shared gate or fence. It is particularly important that pygmy does, which run a higher risk of needing a caesarean in any case, are not bred until they are at least eighteen months old. Do not put pygmy goats or small-framed does to large-breed males, as serious kidding issues are likely to ensue.

Entire male kids become sexually active at just a few weeks of age, which is why they need to be separated from females by the time they are six weeks old to avoid unwanted pregnancies. Does come into season in autumn with the shorter days, which trigger the onset of oestrus. You can tell if a doe is on heat by the following signs: she will wag her tail rapidly, bleat more frequently, her vulva will become red and swollen and possibly there will be a clear mucous string, she will show a lack of appetite and general change in behaviour.

A doe's urine contains pheromones that trumpets to the buck that she is ready to breed; as she squats to urinate, the buck puts his face directly into her urine, and curls his upper lip to test if she's ready. The buck will stay close to the doe's side and lift a front leg tapping at her and this can go on for a day or more until she is fully receptive and in standing heat, when the buck will mount and mate with her. Several matings will happen within an hour.

You need not keep the buck running with the does for the whole breeding season; they can be kept separately and just brought together for the actual mating. For dairy goats there is a particularly good reason for this as the pheromones they produce when a buck is present can make their milk very goaty.

KIDDING AND REARING

Goats have a gestation period of approximately 150 days, ranging from 143 to 157 days. Have the does scanned for pregnancy 80–100 days after mating in order to manage pre-kidding feeding. For the most part you really need to be around at kidding time round the clock, checking things thoroughly at least every two hours, day and night. This should be the key consideration if you have commitments away from your holding. If you work full time you can plan to take your holidays at kidding time and put the buck with the does for a tightly regulated window of opportunity (or use artificial insemination), to ensure that kidding is confined to a manageable period. Far from being a holiday, however, it will be demanding if enjoyable work.

The process and equipment for kidding is the same as for lambing, as is care of the newborn kid: *see* Chapter 10, Sheep.

Goat milking machine.

DAIRYING

A doe will lactate for an average of 284 days, with the peak in milk yield at four to six weeks after kidding. Many goats can be milked for two seasons without getting back into kid.

Dairying equipment can be simple and homely or modern and complex, or a mix of the two, and hygiene is critical. The shelf life of milk is the best gauge as to how hygienic the milking process is, as good hygiene results in longer-lasting milk. It is critical to keep a machine clean: there are some excellent videos online on how to use and clean them.

If you are hand milking you can milk outside in good weather, but keep a clean container with all the items you need close at hand so they don't get put on the ground until after you've finished using them, and then clean them thoroughly before the next use.

Although it's clearly possible to milk a goat standing on the ground, most milkers find a raised stand essential. They are simple to make for even the most basic DIYer. The platform should be around 38–45cm above the floor for ease of milking. Before building one do test out the height at which you'll be comfortable milking when seated in relation to the position of your goat's udder; it's easier to cut down the stand legs than add height. Include a lipped shelf

Goat on upcycled milking stand.

or bowl ring outside the head yoke to keep the feed bowl in place during milking. The stand can also be used when trimming hooves.

It makes the domestic milker's life easier if you give your goats some desirable feed that they only get when on the stand, such as dairy nuts mixed with chopped apple or carrot. For the fidgety goat ensure there is adequate feed to last until their milking is finished.

Hand Milking

If you milk by hand there is minimal capital outlay and less washing involved as you will be dealing with buckets and simple containers rather than dismantling and cleaning a milking machine. Hand milking is also a lot quieter. Goats mostly have teats of a size that accommodate hand milking (which is not always the case with cows). Milking by hand will take longer than machine milking, although this will improve as you become more adept. It can be harder to maintain scrupulous hygiene with hand milking, with dirt and dust more likely to contaminate the milk in an open pail.

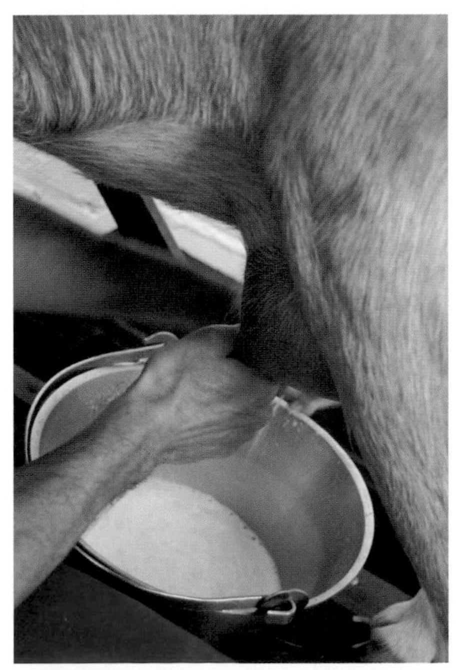

Hand milking into a bucket.

Hand milking can have its problems, including crotchety goats kicking over the bucket, so keep a solid grip on it while they learn to settle to the routine. Start off milking one teat at a time, holding the bucket with your other hand. As the goat relaxes, move to milking both teats simultaneously. If the doe hasn't been milked before she should have plenty of handling pre-kidding, spend time getting used to being on the milking stand, associate being fed her favourite foodstuff with being on the stand, and have her udder touched regularly so that future milking is as stress free as possible.

Avoid pulling on the teats when hand milking as this causes damage. Close your thumb and first finger together at the top of the teat, and gently and firmly bring your other fingers towards your palm to squirt out the milk. Milk both teats at the same time, alternating the squeezing motion with each hand so that first one teat then the other ejects milk into your bucket. Halt milking when it stops flowing steadily, and massage the udder to encourage the doe to let down more milk, then continue milking; you may need to repeat this a couple of times. Stop milking when no more flows.

Hygiene Routine for Milking

- Keep all your equipment spotlessly clean (lying on a shelf in a dusty open barn won't do).
- The hair on the udder, flank and hind legs can be clipped short and the doe brushed to remove any dust, dung or loose hair before putting her on the milking stand.
- Keep yourself clean, clothes and hands. Wear thin nitrile, latex or vinyl disposable gloves. If you have a cut on your hands or arms, clean the wound and cover it with a waterproof dressing before milking, and if you have food poisoning or an infection of any kind get someone else to do the milking.
- Keep your goats and their udders clean using disposable dairy wipes. Avoid using water to clean udders as the dirty liquid will flow on to the teat, potentially contaminating the teat opening. If it is necessary to wash the udder use a dairy disinfectant, and wash and dry the udder and teats thoroughly using a clean paper or cloth towel before continuing.
- Use a strip cup to catch the first three or four squirts of milk from each teat (fore-milking) to check for signs of mastitis (clots or blood) or dirt from the teat. The strip cup has a black strainer that helps you see any abnormal milk. Don't check milk for abnormalities by squirting it on to the floor or your hand or boot, as this can spread mastitis from one goat to another.
- If used, pre-dip should be applied to teats after they have been fore-milked, left for thirty seconds and then dry-wiped, or washed and dried. Use an appropriate pre-dip product and not your post-milking teat disinfectant, as its use as a pre-dip may contaminate the milk.
- Milk into a seamless stainless-steel bucket so there is nowhere for milk or bacteria to collect.
- Strip out the last of the milk from the udder into the strip cup to check again for any signs of mastitis. Post-milking teat-dip disinfectant should be applied as soon as milking is complete while the teat canal is still open.
- Filter the milk immediately to remove any particles of dust, goat hair, skin flakes.
- Chill the milk straightaway and keep it below 3°C. This may be a few degrees cooler than your domestic fridge, but the milk will stay fresher for longer at the lower temperature so you may wish to invest in a small fridge specifically for dairying. For small amounts, initial chilling of milk can be done by standing the filtered milk container in a sink of running cold water.
- Glass is the best storage option for raw milk, and bottles with plastic lids can be thoroughly cleaned in a dishwasher and sterilised as necessary. A small stainless-steel churn is also an option.
- After milking, rinse the stainless-steel bucket with cool water before cleaning it thoroughly

with detergent and hot water. Rinsing with hot water initially just hardens the traces of milk on to the bucket, making it more difficult to clean.

- There are various types of dairy steriliser available (some iodine based, others with hypochlorite), which you can use for cleaning all items too large or inconvenient or inappropriate to sterilise by boiling.

Strip cup.

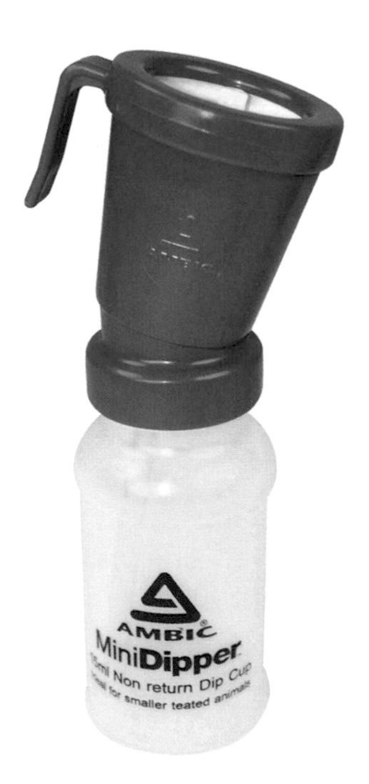

Teat dip cup.

PASTEURISATION

Kept properly chilled, clean raw goat milk should last at least a week, although some goat keepers won't keep it beyond three days; you will have to experiment. By day four a stronger goat taste starts to develop. Pasteurising milk kills off the live bacteria.

To pasteurise, heat the milk to 63°C for thirty minutes, and then cool to under 10°C, stirring throughout the heating and cooling process.

Raw, unpasteurised milk sours rather than rots, and there are numerous ways of using sour milk as a replacement for baking powder, buttermilk, sour cream or yogurt in recipes, including using it as a leavening agent in scones, pancakes, bread and cakes, or in making cottage or ricotta-type cheeses. Quiet handling of the milk (gentle stirring and slow heating) will minimise the goaty taste of the resulting cheese.

Once raw sour milk separates (and smells – goes 'off'), it needs to be thrown away.

Kid-at-Foot Dairying

Keeping calves with their dams and sharing some of the milk for the household has been a smallholder approach for managing house cows since time immemorial. This also works for goat keepers who don't like the idea of removing the kid from the doe prematurely, wish to milk once, not twice a day, and for those who want the option of being able not to milk for a day or two occasionally.

Ethical dairying, as it is sometimes described, has a number of key features. When a dairy animal gives birth, she and her kids/calf are left together to bond for at least a week, with the young receiving all the colostrum and milk. A week after kidding/calving, or a little earlier if the kids/calf struggle to keep up with mum's milk supply, a small amount of milk is taken. For does you can start milking around 200ml, increasing daily in 50–100ml increments to achieve a litre or 1.5ltr per day. Taking more can increase milk production, which in turn increases udder size, which may not be entirely desirable for the domestic milker. A modest udder may be less productive but is less likely to be as disabling as an oversized udder can be for a doe.

When the kids are three weeks old (timescales vary), they are separated from their dams at night, usually kept in the same barn, separated only by a gate or barrier with doe(s) on one side and their kids on the other. In the morning the doe is milked – either fully milked out or an amount left for their kids – then reunited with their kids for the rest of the day.

Kids can be kept with their dams until weaning at eight to ten weeks old, or even later at four or even six months, with the exception of entire bucklings that need to be weaned and separated permanently from females at six weeks to avoid them breeding. Around six weeks into lactation the doe will produce 2.5–3ltr per day (or considerably more, depending on breed, feed and milking system).

If you have young twins or triplets suckling equally on both teats, not milking for a day or two shouldn't have negative effects on milk yield, but don't leave a kid-at-foot dairy doe unmilked for more than that – and as her yield increases, she will need milking daily.

The Home Dairy

The domestic kitchen is perfectly suitable for processing milk into butter, cream, cheese, yogurt, kefir and more for home use (but not for sale to the public). Equipment should be easy to sterilise and keep clean, the milk washed off with cool water immediately after use, then the equipment cleaned with hot water and your chosen dairy disinfectant. If you use wooden kitchen utensils such as scotch hands or spatulas, clean these thoroughly and then pour boiling water over them.

Cheesemaking in a home kitchen.

Saanens and British Alpines.

Drying Off Dairy Does

Most goat keepers will milk a doe for nine to ten months after kidding, then dry her off (to stop producing milk), two to three months before she is due to kid again. The doe has a chance to put on weight, improving body condition lost during the demands of lactation before kidding again.

Three weeks before the planned drying-off date, slowly reduce the amount of energy in the diet by switching entirely to hay, reducing the calorie and nutrient content. After two weeks assess the goat's udder to see how much milk is being produced, and check that the udder is healthy. You can stop milking abruptly to dry off the doe if milk production is now minimal and the udder is sound. If you are concerned about the udder, talk to your vet.

Alternatively a more gradual approach can be adopted: if you milk twice a day, reduce this to once a day. Leaving a little milk in the udder after each milking will also contribute to reducing milk yield. You can then milk every other day, then once every three days, before stopping entirely.

For information on rearing goats for meat, *see* Chapter 14.

For more detail about this fascinating and useful species, *see* my book *Keeping Goats – A Practical Guide*.

12 Cattle

The smallholder dream usually focuses on chickens, a couple of pigs, perhaps a goat or two, or a mini flock of sheep. But cows? Much bigger and maybe intimidating, cows might take you right out of your comfort zone. There's no denying that a cow is a lot bigger than a goat or sheep, though you can get smaller cows of course, such as the Dexter (300–350kg), the Shetland (350–450kg), the Jersey (400–500kg), the traditional Hereford and the Galloway (both averaging 450–550kg). But just as a Great Dane can be gentler and easier to handle than a feisty terrier, you shouldn't equate smallness with docility, or a large scale animal with a challenging temperament.

Certainly some cattle are livelier than an inexperienced smallholder would wish for, but it's all about sourcing well-handled stock, plus having sound handling facilities that keep you safe and the cows secure. Buying a couple of cows and calves that have ranged freely on the moors, never seeing a familiar person from one week to the next and only handled occasionally for TB testing, isn't for the inexperienced. When buying cattle you want to be sure that they come from a herd that is regularly handled, that they don't spook at human interaction, and clearly enjoy having people around them.

Spend time observing how their owner works with them and moves through the herd. I expect to be able to go into a field with my cows safely, and unless I call out to them, that all they'll do is turn to look at me, acknowledge my presence, and then continue grazing or chewing the cud, or perhaps stroll up to say hello. When they are housed for the winter I can move among them without creating a stir, to give them a scratch on the tail head, and they rest quietly as I do so. I don't want any cow shaking its head at me, taking an aggressive stance.

If you have dairy cows to provide milk, to make butter, cream, yogurt and cheese, you'll be milking once or twice daily, and this frequent handling will help you get used to being up close to them. Nevertheless, because of their size you always need to have your wits about you – having even the gentlest cow step unwittingly on your foot is not going to do your toenails any good. We have a small suckler herd, a group of cows that suckle and rear their own young, kept for beef. Because cows are long-lived and rear a calf every year, you might have your breeding females with you for fifteen years or more. Just think of the relationship and bond you can create over that time, how you can rely on them knowing your routine and expectations, and how you in turn appreciate and understand their behaviour and needs.

However, you don't need to breed from cows to rear your own beef. You can provide the household with home-reared prime steak, roast fore-rib, all the mince you could ask for, plus pot roast brisket, steak and kidney and more by purchasing a couple of weaned steers (castrated bull calves) and rearing them on to meat weight at around thirty months; this will give you the experience of cattle keeping with minor hassle. Steers make biddable youngsters: they are lively, curious, friendly creatures, and not prone to the three-weekly hormonal cycle of heifers.

SUITABLE BREEDS FOR THE SMALLHOLDER

Traditional native breeds are the best bet for the smallholder: they are low maintenance, rugged, and thrive on nothing but pasture, whether as grass or grass converted into forage. Choosing between horned or polled breeds is a personal preference;

Herefords with tombstone feeder for horned cattle.

Round bale feeder for polled cattle – Red Ruby Devons.

I would tend to avoid long, pointy horns. Although understanding the attraction of Highland and Longhorn cattle, horns add a further dimension of risk into your cattle keeping. If you choose horned cattle your feeders and crush will need to be of the appropriate design. Some breeds have both polled and horned cattle (such as Red Ruby Devons), and many cattle, although naturally horned, are disbudded shortly after birth to stop the horns from developing.

You have plenty of choice of purebred animals, plus all manner of cross breeds. The descriptions below relate to the traditional types rather than any 'improved' modern versions. Weights are obviously highly variable between individuals, even within a specific breed.

Traditional Breeds of Cattle

Types	Description	
Beef breeds		
Aberdeen Angus	Polled, hardy, long-lived, mostly all black with some all red. Cows 550kg, bulls 850kg.	*Aberdeen Angus bull.*
Beef Shorthorn	Hardy, docile and milky. Horned. Coloured red, white or roan. Originally dual purpose, now split into separate dairy and beef breeds. Cows 500–700kg, bulls 800–1,000kg.	*Shorthorn bull.*

Types	Description	
British White	White with black (or red) points on the muzzle, ears, feet and teats; polled. Milky, medium sized, cows 550–700kg, bulls 900–1,100kg.	 British White.
Devon	Rich red ruby coat with white tail-end. Do well on poor grazing, making them excellent conservation grazers. Polled and horned types. Originally dual purpose, so milky. Long lived, docile, medium sized, cows 650–800kg, bulls 850–1,000kg.	 Devon.
Galloway	Old, hardy, long-lived and milky. Polled. Used for conservation grazing. Black, dun, red, riggit and belted variations with thick, curly coat. Cows 450–650kg, bulls 750–900kg.	 Belted Galloway.
Hereford	Red with white stripes along the back and underside of the belly, and white face. Horned. Hereford bulls pass on the white face to crossbred offspring. Used for conservation grazing. Long-lived, docile, medium-sized, cows 450–550kg, bulls 750–850kg.	 Hereford.
Highland	Distinctive long coat and long horns, commonly red, they can be yellow, brindle, dun, white or black. Thrive in harsh conditions, long-lived and seldom housed. Lean beef. Conservation grazers. Cows 450kg, bulls 650kg.	 Highland.

Types	Description	
Lincoln Red	Large, polled breed, long-lived, docile and hardy. Deep cherry red. Originally dual purpose, so a milky breed. Rare. Cows 700–750kg, bulls 1,000–1,100kg.	 *Lincoln Red (David Bingham).*
Longhorn	Docile, with long, curving horns; colours range from terracotta red, rich red, roan, red brindle and plum brindle, with variable amounts of white; invariably a long white line down the spine, known as finch back. Cows 650–850kg, bulls 1,000–1,200kg.	 *Longhorn.*
Luing	Pronounced 'ling', hardy breed, three-quarters Beef Shorthorn, one-quarter Highland. Red brown or dun coloured, medium-sized, cows 500kg, bulls 950kg.	 *Luing.*
South Devon	Largest native British breed. Light to medium red curly coat. Mostly horned but some polled strains. Milky and quiet. Cows 800+kg, bulls 1,200–1,500kg.	 *South Devon.*

Types	Description	
Sussex	Hardy, docile, rich red-brown, with creamy white tail switch. Horned. Medium-sized, cows 650–700kg, bulls 1,000kg.	 Sussex.
Whitebred Shorthorn	Docile, long-lived, outer coat of creamy/white soft hair and thick undercoat, small horns. Crossed with Galloways to create the Blue Grey. Rare. Medium size, cows 550–600kg, bulls 800–900kg.	 Whitebred Shorthorn.
White Park	Ancient, white with black points and long curving horns. Used for conservation grazing. Rare. Medium sized, cows 600kg, bulls 900kg.	 White Park.
Dual purpose breeds suitable for dairy and beef		
Albion	Blue-roan, white or mainly black with little white. Believed to come from the Welsh Black crossed with a White Dairy Shorthorn. Rare. Cows 500–550kg, bulls 800kg.	 Albion.

Types	Description	
Dexter	The smallest European cattle breed in black, red and dun. Polled and horned strains. Short-legged and non-short-legged versions (care must be taken when breeding short-legged bulls to short-legged females as this can result in aborted calves). Cows 300kg, bulls 450kg.	 *Dexter.*
Irish Moiled	Long-lived, docile and polled, red or roan with a varying amount of white. Used for conservation grazing. Rare. Cows 600kg, bulls 800kg.	 *Irish Moiled.*
Northern Dairy Shorthorn	Medium-sized, red, white, roan or red and white with small, upswept horns. Long-lived and milky. Rare. Cows 550kg, bulls 750kg.	 *Northern Dairy Shorthorn.*
Red Poll	Hardy, medium-sized, deep red with white tail-end. Used for conservation grazing. Polled. Cows 525kg, bulls 750kg.	 *Red Poll.*

Types	Description	
Shetland	Originally a dairy breed so are milky, now mostly kept for beef. Usually black or black and white but also red, dun, grey, brown and brindle. Small, upswept horns. Rare. Small size, cows 350–450kg, bulls 550–600kg.	 *Shetland.*
Welsh Black	Hardy with black, thick long coat. Horned (there are polled strains). Now mostly used for beef. Cows 600–800kg, bulls 950–1,100kg.	 *Welsh Black.*
Dairy breeds		
Ayrshire	Colouring any shade of red or brown and white. Horned. Produces large quantities of quality milk from forage. Medium-sized, cows 450–600kg, bulls 650–900kg.	 *Ayrshire.*
Dairy Shorthorn	Docile and long-lived, producing calves up to seventeen years. Traditionally red, white or roan. Rare. Cows 550–650kg, bulls significantly more.	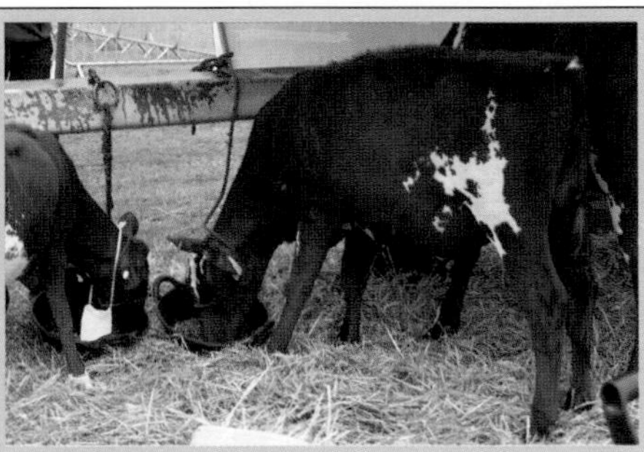 *Dairy Shorthorn.*

Types	Description	
Gloucester	Black brown with a white stripe along the back, white tail and underbelly. Horned. Their milk is ideal for cheesemaking. Rare. Medium size, cows 600kg, bulls 800kg.	 *Gloucester.*
Guernsey	Colouring is fawn and white. Horned. Produces rich, quality milk. Medium size, cows 450–500kg, bulls 600–700kg.	 *Guernsey.*
Jersey	Produces rich quality milk, and a high stocking rate of up to 2.5 animals per acre on quality pasture. Colouring mostly pale from cream to fawn but can be much darker. Hard black hooves. A small breed, cows 350–400kg, bulls 600–700kg.	 *Jersey.*
Kerry	Long-lived, hardy, thrifty small breed useful for boggy, hilly or difficult ground. Black, sometimes with white on the udder. Horned. Slow maturing. Cows 350–450kg.	 *Kerry.*

THE CATTLE KEEPER'S YEAR

The usual 'typical year' caveat is that the calendar of cattle activities will vary depending on your personal preferences, location and land type, noting that cattle can breed year-round. Below is a typical cattle year for us, with a small native beef breed suckler herd in Devon, UK. If you can keep them out over winter it's much easier.

Management Tasks Month by Month

Month	Management tasks
January	The cattle are inside for the winter. Most years a steer will be sent to the butcher.
February	Annual compulsory TB test. Vet also takes blood samples for analysis to test for Johne's, leptospirosis, infectious bovine rhinotracheitis (IBR) and bovine viral diarrhoea (BVD) (these blood tests are by choice). Weigh all cattle. Pregnancy progress checked by the vet. After clear TB tests, any heifers we are selling go to their new homes.
March	Getting closer to turnout time.
April	Cows turned out to grass – time of great glee for us and the cattle! Cowshed thoroughly mucked out, sheds steam-cleaned and disinfected.
May	Cows come inside to calve a week before their due date. Turned back outside once the calf is a few days old, tagged and doing well.
June	Calving continues.
July	Start to AI two-year-old heifers and cows for calving from May onwards the following year. Hay making.
August	AI continues. Hay making.
September	AI continues. Start sending steers to the butcher. This can happen year-round depending when steers are ready to go for beef. Heifers that we don't want to keep or sell for breeding also go for beef.
October	Cowsheds readied to bring the herd inside – sometimes the weather dictates that we need to bring the cows in immediately so the sheds must be clean, beds strawed up, forage on hand, water troughs clean, and any electrics/roof or other infrastructure issues sorted in advance.
November	If we're lucky, the cows won't need to be brought inside until the end of the month, entirely dependent on ground condition and the weather. As soon as the ground is in danger of being poached the cattle come inside for the winter. Each cow has its dorsal strip trimmed the day it comes in and is weighed to monitor progress and for any future medication purposes. The hoof trimmer trims hooves of mature animals (rarely needs doing to animals under twenty-four months). All cows are given a flukicide after housing (timing depends on type of product used). Calves wormed if necessary.
December	The cows are now in, which means daily mucking out, feeding hay/haylage morning and evening. The cows that have been AI'd are pregnancy scanned.

Clipping the dorsal strip with electric clippers to minimise the risk of pneumonia when housed.

FEEDING

Although cattle can be fed on a variety of foodstuffs, pasture, whether as grass in the field or preserved as hay or haylage, is the most natural, sustainable (and cheapest) option, and will result in better tasting beef, and cows with fewer digestive issues. Even if keeping cattle outside the year round, it is likely that you will have to supplement their pasture diet with preserved forage during the winter.

When the grass is in short supply or when cattle are housed, a 650kg cow will eat 12kg of hay per day, or 48kg silage. A very approximate calculation (easy to use and very much on the generous side) is to allow one large bale (approximately 225kg) per adult for each month it is housed, or half a small bale of hay per day. Hay does not need to be eaten in a hurry, but wrapped haylage bales once opened will last only five to seven days before they start to rot; this is indicated by heat, and any warm haylage should not be fed but thrown on the muck heap. If the haylage is wetter than desirable – our preference is to make haylage that is very close to being hay – we will feed hay or straw alongside to ensure that the resulting dung is not liquid.

You can choose to feed cows ad lib, giving them access to a feeder that has a continuously refreshed supply of forage, but we feed housed cattle twice a day, morning and late afternoon, which stops them from gorging and minimises wastage.

We always have a sack of cattle nuts on hand, but rarely offer them to the herd – well fed native breeds really don't need them. Cattle nuts are only used as treats: to tempt a reluctant cow into a trailer or the crush; to cheer them up during/after an injection or slightly stressful activity so they don't worry about being handled subsequently; to encourage them through various fields to where you want them to be; and lastly, for briefly separating mum from calf in order to tag and/or castrate the calf. A cow with twins may need supplementary feeding to ensure she

Devon Ruby cow and her twins.

has adequate nutrition to provide plenty of milk for her multiple offspring.

Feeding a House Cow

A house cow (aka milk or dairy cow) will have greater demands on her than a beef cow, simply because she's producing more milk, and she'll need to eat well if you want her to provide plenty. Ideally her needs should be met by good diverse pasture, supplemented by snacking on accessible hedges. Trying to maximise her output by feeding concentrates can unbalance her rumen, and producing very large quantities of milk puts unnecessary strain on the cow; a smallholder is likely to prefer a long-lived, healthy cow with moderate output than a short-lived one that produces more milk per day than the home can manage. A house cow in milk weighing 450kg can eat as much as 75kg of grass or 12.5kg of hay a day, the same as a 650kg beef cow.

Supplements

Even if cattle are raised entirely on grass, that doesn't mean you shouldn't give supplements to support good health. At turnout in the spring we put out a magnesium mineral bucket to avoid 'staggers', when the lush grass limits the take-up of essential magnesium. If we have a fly problem, we offer a garlic mineral bucket as a first intervention, a non-chemical approach to fly control (plus home-made apple cider vinegar with added citronella or tea-tree essential oil to spray on the cows).

When the cattle are housed they have access to lump rock salt in their mangers, and are given a small amount of seaweed meal as it is full of minerals including iodine and phosphorus, critical for the in-calf/freshly calved cow. If your pasture or forage is missing essential minerals (samples can be tested), these can be offered as a mineral lick or administered via a bolus.

Water

Locate drinkers so that all the cattle, small or large, can reach them. In winter, if there is a danger of water freezing it is helpful to have water troughs big enough to pour in buckets of warm water to sustain the cattle for a few hours; small drinkers may make it easier to clean out if a cow dungs in it, but would not satisfy thirst in freezing temperatures. Some cows play with the water, so place indoor drinkers in an area that drains to avoid a build-up of moisture. If you are using feed barriers, place drinkers outside these so the cows put

Rock salt.

Creating hard standing for a new water trough using crushed stone.

HOUSING

Because our farm is largely clay and rainfall in Devon is high, making the soil heavy, we have to house the cows for five months of the year, usually from mid-November to early April, to prevent them trashing the ground. Housing means that in addition to making plenty of hay or haylage to feed the cows, straw has to be bought in for their bedding, and we choose to muck out daily to keep them clean and the cowshed as healthy an environment as possible. This does not mean a complete clean-out: we remove any cow pats off the straw bed, completely clear the concreted area of muck and kicked about bedding, and scatter fresh straw on the bed. As well as using straw as bedding material, the cows will enjoy eating good quality straw, and frequently eat some immediately after fresh is scattered.

Cattle Water Consumption

Stock Type	Estimated drinking water allowances (litre/head/day)
Early lactation suckler cows	50–70
Mid-to-late lactation suckler cows	40–60
Dry suckler cows	14–40
Stock bull	30–80
Growing cattle	15–50
Finishing cattle	25–75
Dairy cattle in milk	70–150

their heads through to drink, minimising the poo-in-water risk. Create an apron of hard standing for outdoor water troughs as it helps minimise liver fluke and poaching.

Cattle water consumption varies widely according to weight, weather, whether the cow is lactating (and how much), and the moisture content of their feed. Dairy cattle producing high yields of milk can drink up to 5ltr of water for every litre of milk they produce.

Cowshed.

Cleaning out the cowshed: a daily task.

If you live on granite, chalk or other well drained ground you will have a much easier life, keeping cows out year-round, providing an open shelter if needed and forage in the winter. It's not the cattle that aren't hardy enough to be outwintered, it's the damage they would do to the ground that is the issue.

Space allowances for cattle housing from the Soil Association is helpful. Cattle with horns should have additional space and be housed separately from those without.

Our two cowsheds are 6 × 13.7m each and house up to six Devon Ruby cows plus their calves, or eight or nine youngstock per shed, which equates to a generous 14sq m per cow and calf (each cow weighs 650–750kg). Of course it is more spacious at the beginning of winter, but this changes as the calves grow.

All cattle housing must provide a comfortable, clean, dry lying space on a solid base. The loafing area is best not concreted, but preferably an earth floor tamped with lime, or of crushed limestone to facilitate drainage of urine; keeping housing and the atmosphere in it dry is crucial to cattle health. The area where they stand to feed and drink is best concreted, with a rough surface to stop them slipping. You don't need to worry about feeding space when cattle are outside: if you have a large bale feeder kept full, cows will help themselves ad lib. Indoors, about 75cm eating space per cow is a good average, with more for horned cattle.

HANDLING FACILITIES

Unless you have a house cow used to being tied up every day as you milk her, you'll need handling facilities for cattle. Your vet will also expect you to have a cattle crush when TB testing, taking blood samples, checking for pregnancy and so on, and you will need it to keep yourself safe when administering medication, whether by mouth or injection (some medications are available as pour-ons, which are easy to apply out in the field if your cow will stand still for you). A crush is also the most secure way to express milk from a newly calved cow if she has over-engorged teats that her calf can't get its mouth around. If you can, buy a crush with a head scoop, which will make things easier if you have to replace a cow's ear tag, administer medication orally, or attend to any head or eye issues. Budget allowing, a weigh platform under the crush is invaluable.

Space Allowances for Cattle Housing

Breeding and fattening cattle (liveweight kg)	Lying area (sq m per head)	Additional area (sq m per head)	Total available area (sq m per head)
Up to 100	1.5	1.1	2.6
Up to 200	2.5	1.9	4.4
Up to 350	4.0	3	7
350–500kg	5.0	3.7	8.7
Above 500kg	1sq m per 100kg (minimum 5sq m)	0.75sq m per 100kg (minimum 3.7sq m)	1.75sq m per 100kg (minimum 8.7sq m)
Dairy cows	6	4.5	10.5

Cattle crush with kiosk and weigh platform.

Milking newly calved cow in the crush.

Cattle race.

A secure, well-constructed race through which you can run your cows into the crush is very helpful, and our vet practice requires this. We created our own using the wall of the cowshed as one side, and built the other wall out of railways sleepers and second-hand motorway crash barriers, with a concrete floor. The kiosk behind the crush was a small investment in comparison to the cost of the gates, crush and cowshed improvements, and we wouldn't be without it: it enables you, the AI technician, the vet and the hoof trimmer to stand behind the cow and do whatever is necessary in safety without the next cow in the race getting in the way.

The more gates you have, the easier it is to handle your cows; separating them out for treatment, calving and so on is so much easier if you can segment your barn in this way. Socketed posts are excellent as you can install the gates as necessary for winter housing or handling, and at other times remove them and free up the area for other purposes. As well as socketed posts in the cow sheds we created a corral with them; these can be lifted out when they are in the way or not needed, and makes space as flexible as possible.

Hoof Trimming

Different breeds of cattle have varying foot qualities; black hooves are frequently harder, and breeds such as Galloways rarely if ever need their feet trimming, while our Devons have softer, pale hooves that need attention at least annually. A professional hoof trimmer attends to the cattle once they are housed in winter. It is rare that individuals under twenty-four months need hooves trimming.

Professional hoof trimmer at work.

BREEDING

Artificial Insemination or a Bull?

A bull is a complication too far for many small-scale cattle keepers, but that shouldn't stop you going ahead and having calves born on your holding. A bull is a simple proposition in one sense, in that he knows when the cows are ready to be served and will get on with it. So why not have a bull? Just like any livestock, you don't want daughters being bred to their fathers, which means you'd have to get a new bull every three years (offspring will be ready for mating at two years old). That nice quiet bull you felt confident in handling will have to be sold on if you intend keeping any of his daughters, and you can't guarantee that his replacement will be as docile.

AI technician inseminating one of our cows.

Steer riding a bulling heifer.

Another complication is the need to keep young heifers separate to avoid impregnation too early: heifers can get in calf as early as four months old, which inevitably leads to stillborn calves and serious problems for a mum too young to breed. If you want your calves to stay suckling their mothers and want the mothers to run with the bull at the same time, you'd have to inject the heifers to induce abortion, which is not something I'm happy to contemplate. And what about overwintering? You'll need a strong bull pen to keep him away from young or related females.

A bull is therefore an expensive choice for a smallholder with a few cows, and there's the health risk of bringing in new bulls whether on hire or your own, particularly if you want to maintain a closed herd. Basically, a bull can create plenty of complications.

However, going bull free is not without its own issues: it means relying on artificial insemination (AI). There are AI technicians across the country who will manage this for you. You can source semen straws of every breed and have an account with the AI company, who will store it for you ready for use. This requires you to check your cows regularly to note when they are bulling – when they are in oestrus, come into heat and are receptive to the bull. Cows not in calf come into season every eighteen to twenty-four days, and individuals tend to cycle to their own regular pattern – so my notes tell me that Peaceful comes bulling every twenty-two days, whereas Quartz and Moonstone do so every twenty days.

We wait about sixty days after calving before we put a cow back in calf, which is likely to be her second heat post calving. Heifers start to come into heat from around six months (although can start earlier), but unless they are big for their age, weighing at least 65 per cent of adult weight, wait until they are two years old to get them in calf, so that they have their first calf by the time they are three years of age.

What are the signs that a cow is bulling? As cows come into heat they will move around more, and if they have a calf at foot may not stand to let them suckle. They will also start to ride other cows. At this point it is still too early to have a cow AI'd, but you won't have to wait for long because about twelve hours later she will stand to be mounted by other cows. Her vulva will be swollen, and you may see a clear mucus string or discharge hanging from the vulva or swiped by her tail across her rump or flank. If the cows are inside, bulling activity will churn up their bedding; if you go to muck out one morning and find the bedding in serious disarray, you can guarantee that something has been bulling during the night, even if all seems quiet now.

So, when should you call your AI technician? The time to inseminate is once standing heat is properly underway, and onwards for twenty-four hours. By the time insemination takes place, the cow may no longer be standing, the discharge may be thicker and cloudy, but the cow will still be restless.

You should AI later rather than sooner during a heat, so for the most part, if you see something being mounted one day, arrange the technician for the next day. If you know a cow has been bulling all night and is standing firm to be mounted at dawn, arrange AI for that afternoon or very early the next day. If a cow is standing firm at noon or in the evening, arrange AI for the following morning. The best conception rates are achieved in cows inseminated two to twenty-six hours after first being observed on heat, with the ideal time being twelve hours after onset. If you ring your technician before 10am they guarantee to be with you that day. Note that once you notice minor bleeding from the vulva, you've gone past the point of being able to inseminate successfully.

To observe effectively, you need to spend time checking at the right point of the cycle, so a cursory glance is not enough. Observe for ten to twenty minutes early in the morning and before dusk, and check at various points during the day, too. There are a number of heat-detecting tools, including specialist stickers with paint on that are applied just in front of the tail-head, and will reveal their colour when the cow has been ridden, and you may see the hair on the tail-head rubbed upwards, or mud from other cows' hooves wiped on the flank of the bulling cow.

A real positive of AI is that you can go shopping for the best and most appropriate semen for your cows. You may not be able to justify buying a top-class bull, but you can probably afford to get some pretty fantastic offspring sired by him. Individual semen straws cost from £15 upwards depending on desirability, plus the cost of insemination by an AI technician and its storage until use.

If you need a tight calving pattern because of outside work commitments you can consider hormone treatment to synchronise cycles and calving, although this does have a lower success rate than inseminating to a natural season.

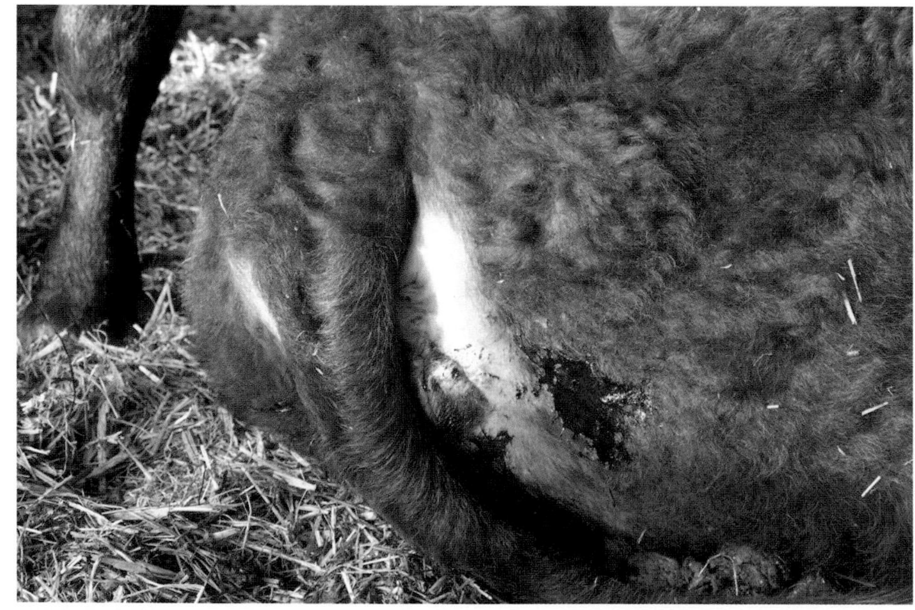

Bleeding after bulling.

CALVING AND REARING

Some people prefer to calve outside, but we bring a cow inside a week or so before her due date (another benefit of AI is that you know exactly when a calf has been conceived). A cow's gestation period is 284 days, very similar to a human, and calving usually takes place up to ten days before/after the due date. When the cows are turned out to grass in the spring, the cowsheds are thoroughly cleaned and disinfected, ready to bring in individuals as they approach calving. Indoor calving means we have light and gates available, and a camera in place so that progress can be observed from a computer or smartphone.

At this late stage of pregnancy cows spend a great deal of time lying down, shifting their weight as the calf moves inside them, and they seem relaxed and unbothered by being back indoors. I use a rubber curry comb to brush the cows in the last week of pregnancy, which they enjoy and which enables them to become familiar with, and unworried by my touch, particularly near the udder and back area, in case any assistance is needed during calving or to help a calf suckle.

For calving presentations *see* the section on lambing presentations in Chapter 10.

Be aware that although twins are not an everyday occurrence, it is not that rare in cattle, and there may be a gap of an hour or so before a second calf is birthed. If twins are a heifer and a bull, the heifer will almost certainly be a freemartin: she will be sterile and unable to have a calf of her own (this sterility

Calving Sequence

Preliminary stage (lasts 2-6 hours or more)	
The cow is uneasy and restless, which might show itself in circling round the pen, stopping for a snatch of hay or drink of water, tail swishing more than usual and udder likely to be engorged. Contractions have started. As this progresses the tail is held out, away from the body. The cervix is dilating. The cow will dung and urinate.	 *Restless and close to calving.*
Second stage (this can take between 2-6 hours or more)	
Strong contractions as the calf's head presses against the cervix; the water bag appears.	 *Waterbag first appears.*
Contractions continue. The dam will get up and down frequently and lick at the liquid dropped from the waterbag.	 *Contractions.*

Contractions continue and intensify.	 *Contracting hard.*
The feet appear: these can pop in and out for some time; if the cow stands at this point they will disappear, but will reappear when she starts having contractions again.	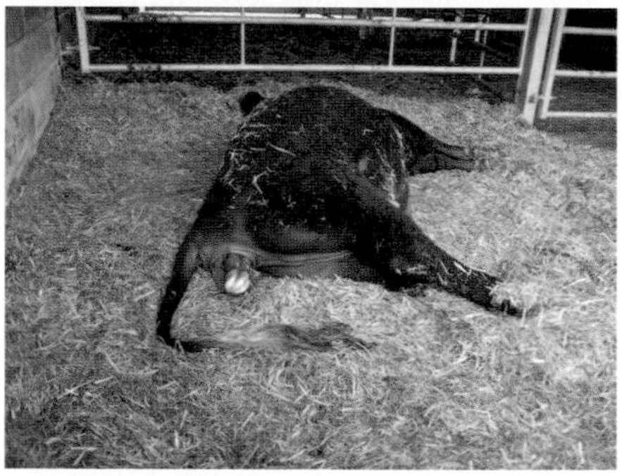 *Feet and nose appear.*
The nose and tongue appear behind the feet, and the cow contracts hard.	 *Feet, nose and tongue visible.*
The head emerges and the calf is born.	 *Head out.*

Within a couple of minutes the dam gets to her feet and starts to lick the calf.	 *Dam licks newborn calf.*
Third stage – passing the afterbirth (can take several hours)	
Do not pull on the afterbirth. In fact you may never see it as the cow is likely to eat it. If it has been retained and hangs from the cow for 48 hours or more, call the vet to remove it.	 *Cow cleansing (afterbirth).*
Within half an hour to a couple of hours the calf should be up and suckling.	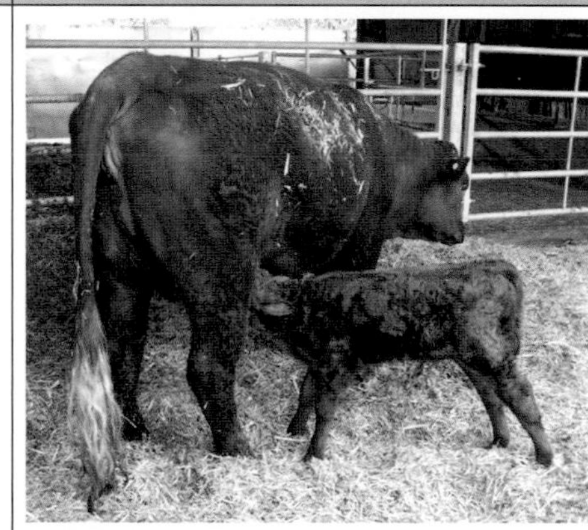 *Calf suckling.*
It's an exhausting business, giving birth and being born – both the cow and calf will rest once the calf has had its first feed.	 *Cow and calf resting.*

can happen in goats although is less common, and is even rarer in sheep).

If you are concerned that calving is not progressing as it should, and don't feel competent to deal with any issues, do not hesitate to call your vet; the value of the cow in economic terms alone will justify the call.

For beef cattle, the mother will do all the necessary rearing required. Your tasks are to ear tag all the calves and castrate the bull calves if required (the elastrator and rubber rings for castrating are the same as for sheep and goats). Although legally you can ear tag beef-breed calves up to twenty days old, I strongly recommend doing it when the calf is no older than 48–72 hours, as long as it is thriving. At this age it spends most of the time snoozing, alongside suckling and finding out how its feet work, and is easy to handle; by seven days a calf is quite feisty; and at twenty days it is much stronger and will be more challenging to handle, and more distressed by the ear-tagging experience, which could upset the mother.

For safety, separate the cow from her calf by offering her some cattle nuts, and closing a secure gate between her and you and the calf. It takes less than a minute to ear tag a calf (and another minute or two to castrate a bull calf), which you can do safely from behind a gate. As soon as you've done, let mum back with her calf; they are ready to go out to grass.

If you calve inside, remove all bedding and clean the area where the cow has calved, spreading hydrated lime or dairy-cow cubicle bedding powder to disinfect the ground/floor area before bringing in another cow for calving.

WEANING

We used to wean calves from their dams at ten months, believing we were allowing the cow to regain her strength and dry up her milk before she calved again. This stressed both calf and cow, so some years ago we stopped weaning entirely. The dam simply stopped allowing her calf to suckle a month or two prior to her next calving, but they continued to graze contentedly alongside each other. By the time we brought the cow inside for calving, her yearling calf was entirely unbothered by the separation and on a 100 per cent grass diet, and the cow's colostrum developed just fine in time for the new calf.

There are nose rings available to prevent suckling – they are clipped into the older calf's nostrils for a week or so and have spikes on that make it uncomfortable for the cow so she resists being suckled – but we have never found these necessary.

THE HOUSE COW

If you keep a dual-purpose or dairy cow to provide the household with milk, you will need to get her used to being milked by hand or machine. For details about milking and dairying, and for keeping a calf with its dam while dairying, *see* Chapter 11 on goats, as the process is the same.

REARING FOR BEEF

And what about rearing your own beef? You might not have the grazing space to keep a small herd of breeding cows, or are nervous of managing a bull or the complexities of AI, but you could buy a couple of weaned beef steers and take them to meat weight. If you want cattle on the holding for the shortest possible amount of time, buy store cattle (fifteen to eighteen months), which means they are approaching meat weight but require anything from a few months to a year to finish.

Alternatively you could buy in and rear calves from a few days old that come from dairy herds, ideally ones that have a beef-breed sire. This will require some fairly full-on care, and a draught-free, well-ventilated shed; the calves will need feeding three or four times a day on replacement milk (the milk is not cheap, even though the calves would cost a lot less than buying older, weaned animals), plus hay and hard feed. Calves can be weaned off milk from around eight weeks if they are doing well on hay and hard feed. If you want rose veal, the calf should be reared until eight to twelve months.

We take our entirely grass-fed Devon Ruby beef steers to thirty months (the heifers to twenty-four months so they don't put on too much fat), which provides 175–220kg of beef depending on conformation – and from the mince to the rib-on-the-bone celebration roast, it is simply exquisite.

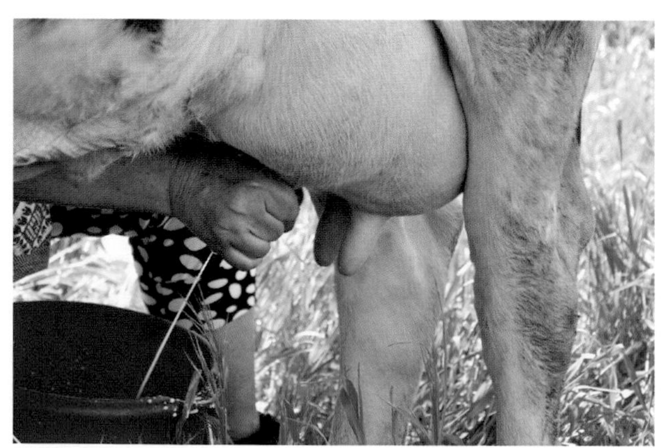

Milking by hand.

Cow and calf dairying.

13 Health and Welfare

I am not a vet, and the information in this chapter is given as guidance only. If you're out of your depth with regard to diagnosis or treatment, which is likely for all but the most experienced smallholder, please do not refer to the notoriously unreliable Dr Facebook – nor to any other social media platforms – which supplies a mix of contradictory, illegal and sometimes dangerous disinformation. Even the best information (and there's plenty of that too, from knowledgeable smallholders, farmers and vets) can get lost in the wash of nonsense on amateur forums.

Instead, please refer to your vet and expert sources: with their help you will build expertise and be able to determine when treatment is suitable (effective and/or economic), and if humane dispatch is the best route, either to save stock from suffering or to maintain the health of the rest of the flock/herd. Your vet is also essential for prescribing any prescription-only medicines (POMs).

Note the withdrawal period for meat/milk/eggs (how many days must pass before it is safe to eat these products) *before* using any medication, as this may affect your decision on timings and the type of medication you use. Some medications, in particular antibiotics and painkilling anti-inflammatories, are only available from a vet; others can be bought from your agricultural merchant. For the most part, specific medications are not named here as they change over time. Please do not give human medications such as painkillers to livestock because they can be fatal in some species; they should only ever be given if directed by the vet.

Normal Temperatures

	Normal Fahrenheit	Normal Celsius
Beef cow	98-102.4	36.7-39.1
Calves	102-102.5	38.9-39.2
Dairy cow	100.4-102.8	38-39.3
Sheep	100.9-103.8	38.3-39.9
Lambs	102-104	39-40
Pigs	101.6-103.6	38.7-39.8
Piglets	102.5	39
Goats	101.3-103.5	38.5-39.7
Kids	101.8-104.5	38.7-40.2
Chickens	105-109.4	40.6-43
Ducks and geese	107.5	41.9

TAKING TEMPERATURES

If an animal is slightly off colour, taking a rectal temperature can quickly indicate if something is seriously wrong. Using a digital thermometer is easy and you should keep one in your livestock first-aid kit. Batteries in digital thermometers are not replaceable, so check your thermometer is working from time to time (in particular in the lead-up to births). Turn it on and insert it gently into the animal's rectum, then tilt it slightly left or right so that it takes the temperature of the rectal wall, not of the faeces. Wait for the beep, remove it, and take the reading. Clean it thoroughly before storing for future use.

KEEPING YOUR LIVESTOCK HEALTHY

Having healthy livestock is best achieved by sourcing healthy stock to begin with, and preventing disease wherever possible. The cornerstones to preventing disease are good hygiene, providing a nutritious and appropriate diet and clean water, avoiding pitfalls (such as incorrect or inappropriate administering of medication), carrying out daily routine checks, timely treatment when needed, quarantining as appropriate, being active in your own never-ending education, and the support of a good vet. In addition, store feed securely so that livestock can't help themselves, throw away any food that goes mouldy, and keep on top of pest and vermin infestations.

Make sure there are no items around that can cause livestock harm: loops of baler twine, rubbish, sharp edges, loose fencing wire, broken glass, poisonous substances such as weedkiller or rat poison – or dead rodents that have been poisoned, to name but a few.

Pneumonia is a particular risk for sheep, goats and cattle when there are hot days and cold nights, so be alert to their wellbeing at these times, and also when housed in damp conditions with poor ventilation.

Signs of Good Health/Ill Health

Healthy livestock are in good body condition, eating well and have a clean mouth, pink gums and inner eyelids, walk soundly, breath evenly, have a good coat, bright eyes, are alert and inquisitive, and deliver faeces of a good consistency and colour for the species. Signs of poor health include loss of appetite, underweight condition, coughing, itching and rubbing, loss of coat or feather, heavy, quick or

Healthy cow pat.

Normal young calf poo – yes, it's yellow!

raspy breathing, cloudy eyes, pale inner eyelids and gums, stained mouth area, stained anal area or scouring, hunched stance, lameness, swellings, abscesses, and so on.

The role of the smallholder is to appreciate the signs of good health so that you spot anything untoward in the earliest possible stage so that it is easier to treat and causes the least stress and pain for the animal. Although not entirely failsafe, a quiet group of animals is generally content; any unexpected bleating, squawking or mooing sends me off on an investigatory tour – although that doesn't include the excitable hen announcing proudly to the world that she's just laid an egg.

This chapter covers the primary issues of which smallholders should be aware; there are many more diseases and conditions that will be familiar to your vet.

Developing a Health Plan with Your Vet

A vet will be keen to work in partnership with you, taking time to understand the level of your expertise, so they can bring their own extensive knowledge and experience most fruitfully to the relationship to the benefit of your animals. They will help ensure the best health of your livestock, and are calmly on hand to help you through emergencies. Many vets include a complimentary initial visit to your holding to find out about the livestock you have.

Few medications are licensed for goats, as the dosage required is often higher than that indicated for sheep, and more frequent treatment may be needed; your vet will therefore have to advise and prescribe off-licence treatments.

Be active in building your own knowledge: reading books, going on courses, and accessing some of the excellent online resources including the Moredun Research Institute, Goat Veterinary Society, National Animals Disease Information Service and the Agriculture and Horticulture Development Board; this means that you won't be bombarding your vet with generic questions. Even so, most vets are happy to answer phone and email queries when you have a question that you can't resolve but which doesn't require a visit. Increasing numbers of veterinary practices run training days for their clients, so take advantage of these; workshops include the safe use of medicines, nutrition, lameness, lambing, calving, and parasite control.

Preparing for a Vet Visit

It saves time if you can describe what the problem is when contacting your vet. Think about the animal's history, its age and condition (whether pregnant, lactating, or with observable symptoms such as a lack of appetite or scouring) and any recent treatments; taking its temperature and relaying this information is also helpful. Let the vet know if the animal is new to your holding or has been in contact with new stock, and tell them if the problem is a one-off or recurring, and whether it is restricted to one animal or not. Make sure the animal(s) in question are in a secure area where the vet is able to handle them, preferably under cover and with light and water available; it's not the vet's job to hurtle round a field trying to catch the patient.

NOTIFIABLE DISEASES

Notifiable diseases are those that you are legally obliged to report to the Animal and Plant Health Agency, even if you only suspect that an animal may be affected. Notifiable diseases can be endemic – those already present in the UK – or exotic, meaning those not normally present in the UK. For most notifiable diseases there are legal powers to cull animals to prevent the spread of disease during an outbreak. There is a long list of notifiable diseases (*see* www.gov.uk), including some that may be familiar such as avian influenza, bovine TB, sheep scab, scrapie and African swine fever.

Weighing your Livestock

Most medication is given according to the weight of the animal, so knowing what that weight is rather than making a wild guess is an important part of healthcare. Weigh scales are not cheap but can be purchased second-hand at farm sales – in which case give them a thorough clean with an appropriate farm disinfectant before use. If you don't have livestock weigh scales and have small or light young animals to weigh, use bathroom scales, get on them and take the reading, pick up the animal,

Second-hand weigh crate for sheep, goats and young pigs.

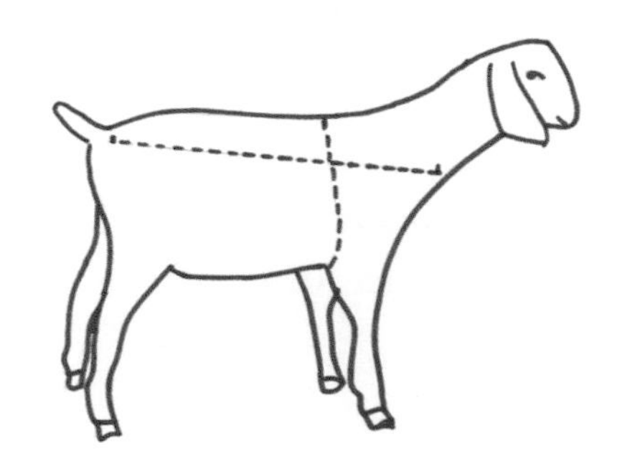

Measuring goats, sheep and cattle to calculate their weight.

take another reading and subtract one weight from the other, and there you have it. Obviously this won't work for cattle, adult mammals or hefty youngstock, in which case a tape measure can help you make a reasonably good calculation of weight. Specific livestock weigh tapes are available that do the calculations for you, but using the calculation below will give you a sound result.

By taking the following measurements you will be able to calculate the weight of your cattle, sheep and goats:

- Measure the circumference (known as the heart girth) of the animal around the body behind the shoulder in inches.
- Then measure the length of the body from the front of the shoulder to the pin bone at the rear in inches.
- Take the first measurement and multiply it by itself, then multiply that by the second measurement. Divide the whole by 300, which will give you the weight in pounds.
- Now you have the weight in pounds you can convert it to kilos by dividing the number of pounds by 2.2.

For measuring pigs to assess their weight, *see* Chapter 9.

Hygiene

There are many aspects to hygiene, of course, mixed with the practicalities of the smallholding being a place where animals and hopefully nature thrive. There will be faeces of all kinds, bodily fluids, and good old dirt. I will never invest in a poo-picker for pasture – faeces from ruminants and poultry deliver nutrients that are critical to soil and insects – but this does not mean allowing livestock areas to become too thickly covered in dung. Clean out poultry houses when needed, which might be weekly, or more or less depending on the weather and the relative size of hut to the number of birds being kept. After calving, lambing, kidding and farrowing, clean areas scrupulously before bringing other animals, particularly newborns, into the area.

Keep yourself healthy by washing your hands after handling livestock, and if you are looking after a poorly animal, attend to it last to avoid transferring muck or other infected material from poorly animals to healthy ones. Don't allow dogs to eat placenta or bits of deadstock before the knacker removes them. Keep drinkers and feeders clean; we're not talking about maintaining the same standards as in your kitchen, but don't allow mouldy food to accumulate, and check that drinkers haven't been contaminated with droppings; if they have, clean them out thoroughly before refilling.

Testing for Diseases and Parasites

Testing can be carried out for a number of key diseases and parasites to understand the health status of your livestock, and any stock you wish to bring in.

Faecal Egg Counts (FECs)

There are faecal sampling tests for intestinal parasites, for all species. It's simple and affordable to test fresh poo samples to identify if they are carrying a worm burden that needs treatment. This ensures that you don't use wormers unnecessarily, or use ones that have no effect on the worms identified. Your vet can provide this service, or there are laboratories and agricultural merchants that take posted samples. Tests can check for a range of parasites including roundworm, stomach hairworm,

Faecal sampling packs.

nematodirus, tapeworm, ostertagia, coccidia, lungworm and liver fluke.

Testing for Tuberculosis

If you keep cattle they must be routinely tested for bovine TB, the frequency (from six monthly to every four years) depending on where you are in the UK, and if the area is a TB hotspot or not. The Animal and Plant Health Agency (APHA) will write to you to explain the type of test you need to arrange, and when; there is no fee for routine TB tests. There is no statutory testing for TB in non-bovines in Great Britain, and the testing of goats, sheep, pigs, farmed deer and camelids is only carried out where infection has been confirmed in a herd, if there is suspicion of infection in a herd, or if private testing is requested by the owner. Confirmation of a positive test result incurs herd movement restrictions, slaughter of livestock that are reactors, and further compulsory testing.

Testing for CAE/CAEV (Caprine Arthritis Encephalitis)

CAE is a viral disease of goats spread via colostrum or milk from infected does, through mating, and by shared equipment such as drenching guns and syringes. CAE can be relatively mild and easily missed in the early stages, but ultimately causes acute loss of condition, lameness, painful knee-joint arthritis, lung infections and mastitis. The effect on the herd is devastating, and goats should be blood tested annually as apparently healthy individuals may infect the rest of the herd. Goats showing clinical signs should be culled.

Testing for CLA (Caseous Lymphadenitis)

CLA is a chronic contagious bacterial disease of sheep and goats that causes abscesses in the lymph nodes in various parts of the body, in particular at the base of the ear, below the jaw, in front of and behind the hind leg, and on the shoulder. Not all abscesses are the result of CLA. Livestock don't appear to be ill until the herd/flock is heavily infected. CLA is confirmed by lab tests of the contents of abscesses.

Infection is spread by the pus from burst abscesses, and any individual with abscesses should be isolated from others until there is a diagnosis; the pus survives to be infectious on surfaces such as walls, hurdles, feed and drinking equipment. Control external parasites to minimise scratching and rubbing that might break the skin, making other individuals vulnerable. There is no treatment, and culling may be necessary.

Testing for Johne's Disease (Bottle-jaw)

Johne's (pronounced yo-nees) is a bacterial disease of adult cattle, sheep, goats, deer and rabbits. Signs include gradual weight loss, and a swelling under the jaw, with scours or soft faeces seen in the final stages. Livestock can be blood tested, but there is no cure or treatment.

Testing for Maedi Visna (MV)

Maedi visna is an infectious, incurable viral disease that is slow to develop and causes wasting and organ degeneration and is ultimately fatal; it cannot be detected early in an animal's life. The virus that causes caprine arthritis encephalitis (CAE) is very similar. Some sheep-breed societies in particular encourage MV accreditation, and there are separated MV-accredited sections at various agricultural shows and livestock auctions.

High Herd Health Tests for Cattle

Blood can be taken from cattle to check whether they suffer from leptospirosis, neospora, Johne's, bovine viral diarrhoea (BVD) and infectious bovine rhinotracheitis (IBR). A single sample can be used for all these tests, and can help in ensuring you do not bring infected cattle on to your holding.

EXTERNAL PARASITES

Lice, Ticks, Mites and Nasal Bots

Lice, ticks, mites and nasal bots can be treated with a topical product applied to the skin, or via injection, and if found in a herd/flock, all animals that are kept together should be treated, as ectoparasites don't restrict their activity to just one animal. Apart from itching, anaemia can be caused by sucking lice (lice charmingly come in two types – sucking and biting), and lice and ticks can be seen by the naked eye. Small numbers of individual ticks can be removed with tick removers such as are used for dogs.

The dreaded red mite is the bane of chicken keepers. Birds that are sitting on a clutch of eggs are more prone to infestation, as is any poorly bird. Red mite lives in cracks in the bird hut (which is why ship-lap is unsuitable for poultry housing) or under roofing felt, while Northern mite lives permanently on the bird. Putting double-sided sticky tape on the ends of perches will quickly reveal any problem (and trap some of the mites). Keep your huts clean with a farm disinfectant and insecticide. If birds are permanently itchy or losing condition, ask your vet to prescribe a pour-on (apply to the back of the neck or under the wing) or medication for inclusion in their drinking water.

Scaley leg is caused by a mite that lives under the scales on chickens' legs, making the scales rise and go crusty. Regularly clean and disinfect housing and perches, and a simple but effective treatment for the birds is to smear petroleum jelly (the stuff we use on our lips, but from a separate pot please!) thickly on their legs, which both soothes the skin and suffocates the mite. Repeat weekly over a four-week period. With severe cases of scaley leg a bird can lose a toe or part of a foot.

Flystrike

Flystrike is a serious risk when the weather is warm and wet, and sheep, goat fibre breeds, hairy cattle such as the Highland, rabbits and chickens and livestock with dirty bottoms (faecal clumps or urine staining) or any wound are at particular risk. It can affect perfectly clean livestock too. Flies lay eggs on the host, and the eggs hatch into maggots, invading the living tissue, while livestock with horns can be attacked by head flies where the horns emerge.

Flystrike preventative treatments for sheep.

In hot, humid weather stock must be checked twice daily for signs of flystrike. Look for flies gathered round livestock, sweaty patches on the flanks, back or by the tail, itchiness, rubbing, biting at themselves, looking generally out of sorts, lameness (look for maggots between hoof claws), and individuals keeping away from the group – and on close inspection fly eggs, or at a further stage – to be avoided whenever possible – maggots.

Strike can develop very quickly, with maggots appearing within twelve hours of eggs being laid. Fly eggs and maggots need to be dealt with immediately: once spotted, clip away any wool, hair or feather from the affected area, check between foot claws, armpits and groin, remove all eggs and maggots, and treat with antibacterial or antibiotic spray depending on the seriousness of the wound (if any). Maggot oil with citronella is soothing and a fly preventative.

There are very effective insecticides available to use as a preventative and/or treatment depending on the product chosen, and if an individual is found to have flystrike the whole flock should be sprayed with a preventative. There are recipes for creating gentle home-made fly prevention treatments (for example cider vinegar with added tea-tree oil and citronella), but using these has to be hand in glove with close observation in case more aggressive preventative treatment with a chemical is required.

The risk of flystrike can be reduced by implementing the following measures:

- Keeping livestock clean by shearing, dagging and crutching to reduce faecal contamination
- Effective worm control to reduce scouring
- Using fly traps
- Prompt treatment of foot rot and wounds
- Prompt removal of fallen stock
- Avoiding grazing in sheltered or shaded pasture with a lot of trees at high-risk periods as they harbour more flies than exposed pasture
- Avoiding grazing by rivers and open water sources at high-risk periods as they attract flies
- Using preventative treatment at the onset of warm, humid weather

INTERNAL PARASITES

Internal parasites live in the body of the host in the digestive tract, cavities, organs, blood, tissues or cells. The subject of wormers (aka anthelmintics) is complex, and veterinary advice has changed considerably over the years now that so many intestinal worms are resistant to chemical wormers. Depending on the type of wormer purchased, it can be administered as an oral drench, a pour-on, or injected; note that the method of administration is specific to each product, and you cannot inject a pour-on, or use a pour-on as an oral dose, and so on. We prefer to use an oral drench to get straight to where the problem lies. Testing faecal samples is critical in managing worm burdens and determining the use of anthelmintics.

Prolonged use of one wormer increases the production of worms resistant to that particular wormer – but paradoxically, changing wormers too frequently can result in multiple resistance developing, so wormers should be changed no more frequently than annually, with some vets advising that one family of wormer is used until it is no longer effective and only then switched. Although there are many different wormers available, there are only five different families (white, yellow, clear, orange and purple), so when changing do make sure you move to a different family, not just a different make. Orange and purple wormers are particularly useful as quarantine treatment when buying in new livestock.

Healthy adult cattle and sheep should not require worming routinely; only treat vulnerable individuals (those that are not thriving, or are coughing, or scouring) showing worm burdens from FECs. Goats are very susceptible to worms, and unlike sheep and cattle never build up immunity and so need regular FECs across the herd to assess if treatment is required.

Sows should be wormed a week before farrowing, and weaners a few days after weaning. We use a powdered wormer in their feed, first adding a little vegetable oil to their sow nuts, stirring it until the feed is well covered and adding in the wormer and giving it another thorough stirring to coat the pellets.

The most useful online resource for understanding good worming practice is www.scops.org.uk.

For poultry, birds should always feel heavier than expected, so pick them up from time to time to gauge weight and check body condition. If lighter than they should be and if the breastbone feels prominent, they are likely to have intestinal worms and need treating. There may be other underlying problems, but effective worming should be your first step. Either add the appropriate amount of wormer to the bird's feed for a week, or buy feed already containing wormer.

Apple cider vinegar can be given to birds (diluted in water, and always served to birds in a plastic container) as a tonic and to benefit the health of the gut, but it won't cure a worm infestation. Always provide fresh plain water alongside. Practise stock rotation with your birds, giving them access to fresh ground, particularly if you are heavily stocked, to prevent a build-up of worm eggs. Disinfectant licensed against worm eggs, coccidial oocysts, bacteria, fungi and viruses can be used to clean bird houses.

Poultry and pig wormers.

Liver Fluke

Fluke is a parasitic flatworm and a serious issue for ruminants on wet land as it compromises the liver and can be fatal if not treated. Fluke is incredibly hardy and built to survive; ask the lab to include testing for fluke alongside worm analysis when submitting faecal samples. When liver fluke is well developed signs include anaemia, bottle jaw and weight loss.

You can mitigate the risk by not allowing your livestock to graze on wet areas by ponds, ditches and rivers, and putting hard standing under and around outdoor drinkers, but if you live in a wet region there is every chance that you will need to treat against fluke. There are combination worm and flukicide treatments available, but use specific flukicides instead for adults as you do not want to contribute to wormer resistance by administering wormer unnecessarily.

FAMACHA Scoring

FAMACHA scoring is a test for barber pole worms (*Haemonchus contortus*) using the symptoms of anaemia to assess whether individual goats or sheep need treatment. The FAMACHA chart shows different colours of the membrane of the eye, from a dark red (1) indicating no significant anaemia, to a white colour (5) that shows severe anaemia. Just because livestock have nice pink eye membranes does not mean they do not have other worm burdens; dung analysis is also needed.

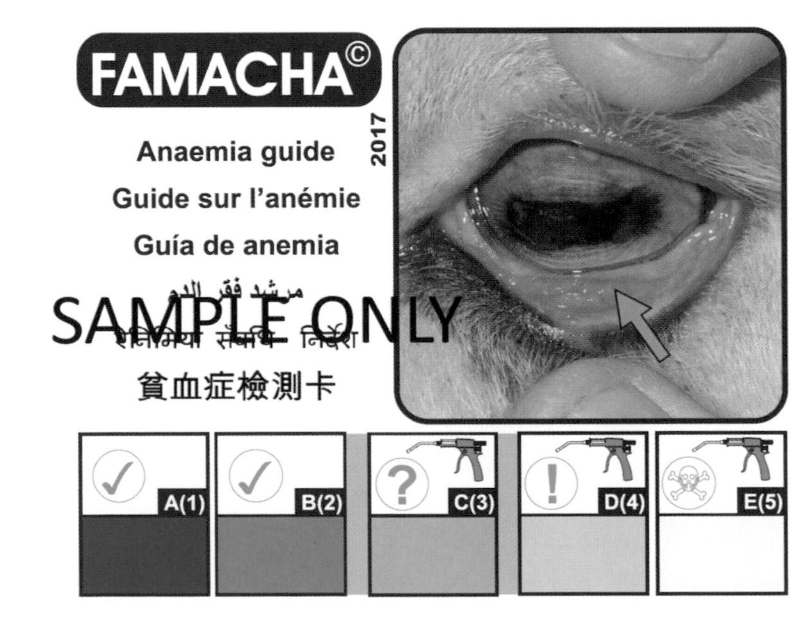

FAMACHA chart.

MINIMISING WORMER USE

To minimise the use of wormers do not overstock your land. Mixed or rotational grazing with other species (cattle, sheep, horses), and access to longer grass, can help minimise worm burdens. It is not appropriate to rotate or co-graze goats with sheep as for the most part the parasites that infect the two species are the same. For goats, the provision of plenty of high-level browse rather than low-level grazing helps: because they remove the tops of taller grasses, and if they have access to trees and hedges plus hay fed above the ground in racks, they are less likely to pick up a worm burden. Grazing on certain plants such as chicory, bird's-foot trefoil and sainfoin reduces the negative effects of worms, and could be helpfully included in a pasture planting plan.

Coccidiosis

Coccidia are microscopic parasites found in the gut. Until immunity develops as the animals mature, youngsters are at particular risk, as coccidia damage the inside lining of their gut causing bloody diarrhoea. There are a number of treatments for coccidiosis, and coccidia can also be tested for in dung samples.

ADMINISTERING MEDICATION

Injections

Depending on the product, injections are given subcutaneously (under the skin), intramuscularly (into the muscle) or intravenously (into the vein); the third option is a procedure to be undertaken by your vet, but a smallholder will be expected to give subcutaneous and intramuscular injections. Most

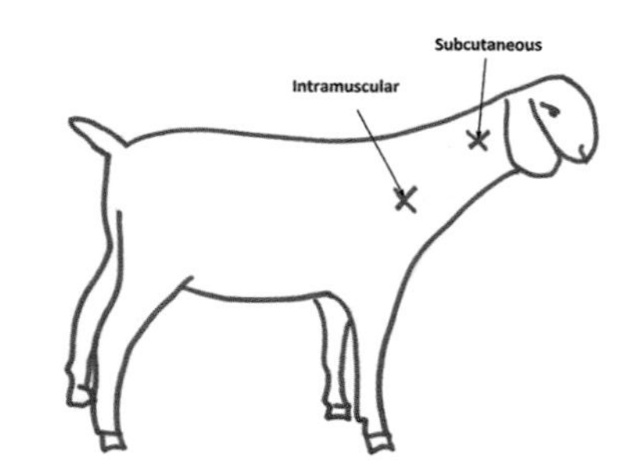

Injection sites.

injections are given into the neck skin or neck muscle, unless the datasheet or vet indicates otherwise. For cattle, injections are given in front of the shoulder or in the rump rather than the neck, to avoid any lumps that may cause confusion at TB test sites.

For a subcutaneous injection, pinch or tent a fold of skin 10–15cm below the ear on the side of the neck where the skin is loose, and inject into the cushion of skin. For intramuscular injection, inject slowly into the neck muscle 10–15cm in front of the shoulder, away from any area of bone and aiming upwards towards the head. For both injections gently massage the area afterwards to disperse the medication.

Needle Gauge and Length for Livestock

Hypodermic needles	Gauge*	Length for subcutaneous injection	Length for intramuscular injection	Cap colour
Poultry	21	10mm	10mm	Green
Goats and sheep				
Kids/lambs under 20kg	21	13mm	13mm	Green
Kids/lambs 20 to 40kg and adults	19	13mm	25mm	Cream
Adults (for thicker solutions)	18	13–16mm	25mm	Pink
Pigs				
Piglets	18 or 20	12–16mm	12–16mm	Pink or yellow
Weaners	16 or 18	12mm	16–18mm	Clear or pink
Finishing pigs	16	18mm	25mm	Clear
Adults	14 or 16	25mm	25–40mm	Pale green or clear
Cattle				
Under 200kg	18	25mm	25mm	Pink
200–600kg (thin solutions)	18	40mm	40mm	Pink
200–600kg (viscous solutions)	16	40mm	40mm	Clear
Adults (for intravenous/large volumes)	14	50mm	50mm	

*The higher the number, the thinner the needle.

Drenching/Dosing Orally

When administering an oral drench, weigh your animals, check the drench gun is delivering the correct dose, and administer it correctly:

- Underestimating weight is a common cause of under-dosing. Either weigh individually and dose accordingly, or select and weigh the biggest in the group to determine the correct dose. The exception is when using yellow wormer, which can be toxic at high levels.
- Calibrate and maintain the drench gun and check its accuracy at regular intervals. Clean with warm soapy water after use, and use a little vegetable oil to keep it lubricated. Calibrate by squirting a single dose from the gun into a small container that you can then withdraw into

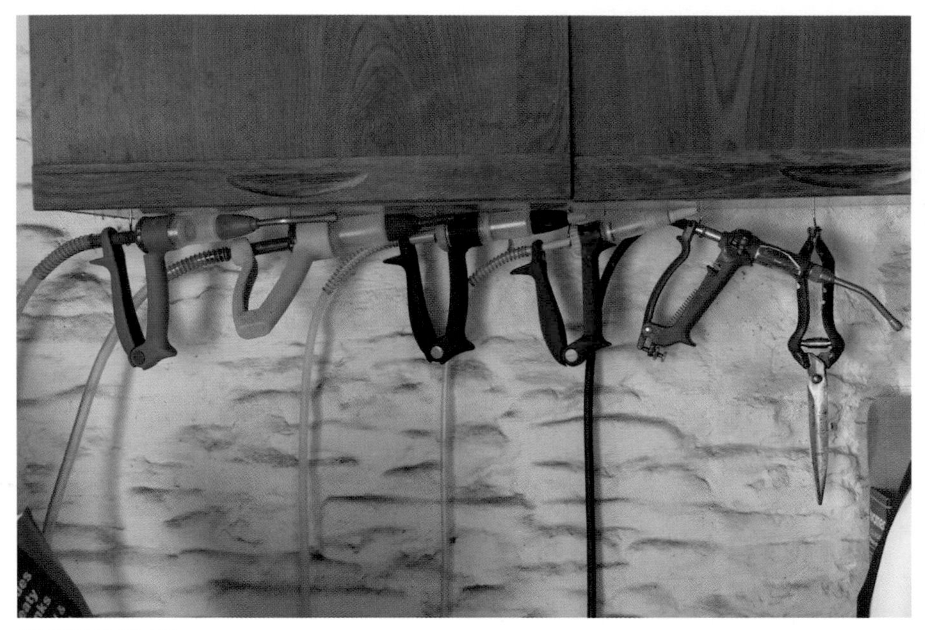

Drench and pour-on guns.

Antibiotic and antiseptic sprays.

a marked syringe to check the correct dosage is being delivered.

- Make sure the animal is properly restrained and can't leap about when you are drenching so they swallow the whole amount. Place a hand under their head and tilt slightly to the side. Slot the nozzle in the gap between the molar and incisor teeth and over the back of the tongue, not just inside the front of the mouth. Pushing the gun down the throat risks damaging the windpipe, and if administered just inside the mouth they may spit out much of the dose.

Topical Application of Medication or Treatments

The topical application of medication or treatments includes sprays and pour-ons. Pour-on medication is probably the simplest of all to administer, but do check how each product should be applied, at what distance from the body, at what site, and with what equipment. Some pour-on bottles come complete with applicator, others require a spray gun. Do not apply when animals are wet or when rain is forecast. Pour-ons are usually applied along the back of an animal (sometimes in a line, sometimes as a spray). Just like oral dosing guns, pour-on/spray guns should be calibrated and the animal weighed to ensure the correct amount is applied.

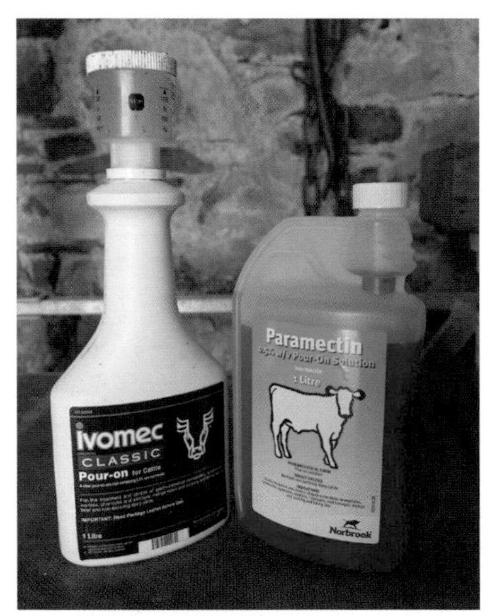

Pour-on medication with built-in applicators.

Antibiotics

Vets cannot prescribe antibiotics unless they visit you every year, and penicillin-based antibiotics is the choice at lambing/kidding/farrowing, for newborns with infection (pneumonia, joint ill, respiratory infections, enteritis), and for dams when you've had to reach inside her to assist with a difficult birth, or to treat mastitis. At other times use a long-acting oxytetracycline-based antibiotic ('long acting' is specified by the letters 'LA' after the brand name, for example Alamycin LA). This is useful when treating animals in the field so that one injection suffices rather than having to catch a sheep, pig or cow to treat for three days running. Oxytetracycline can retard skeletal growth of the foetus if administered during pregnancy, which is why it isn't used for lactating dams or their young.

Antibiotic sprays are for the prevention of infections of superficial traumatic or surgical wounds, and claw/hoof infections, in particular scald, infected wounds, and orf lesions. Antibiotic spray is commonly used unnecessarily in uninfected wounds where an antibacterial spray should be used; we all have a responsibility to minimise the use of antibiotics, focusing its use where needed. Silver aluminium antibacterial spray is recommended for wounds from castrations, disbudding and dehorning, and surgical wounds. For minor wounds such as cuts and scrapes, use an off-the-shelf antiseptic spray often known as violet or purple spray, or a copper/zinc spray.

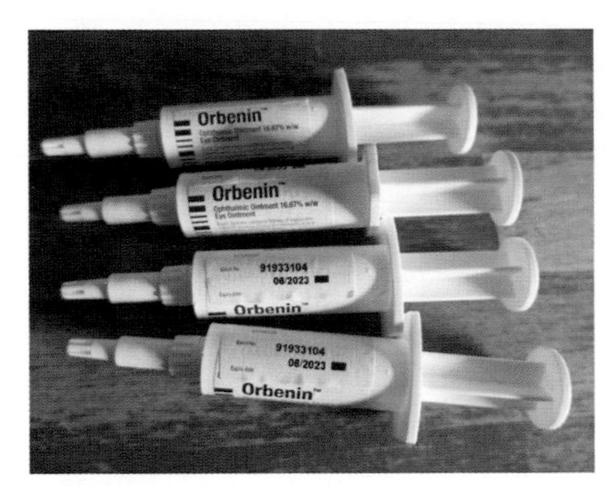

Antibiotic eye cream.

Antibiotic ointments are used for treating eye conditions such as conjunctivitis, New Forest or pink eye in ruminants, and sticky eye and conjunctivitis in birds. Both eyes will require medication as the conditions are highly contagious, and a tiny blob of cream dropped on each eyeball is needed.

Pain Medication/Anti-Inflammatories

Non-steroidal anti-inflammatory drugs (NSAIDs) are used to relieve pain, reduce inflammation, and bring down a high temperature. Often used to relieve symptoms of sprains, arthritis, and other causes of long-term pain, NSAIDs also treat a wide range of infectious conditions including mastitis, respiratory disease, lameness and joint infections, and for castrations and disbudding/dehorning.

LAMENESS AND FOOT CARE

All smallholder species need care taken over their feet, whether we're talking claws, webs, trotters or

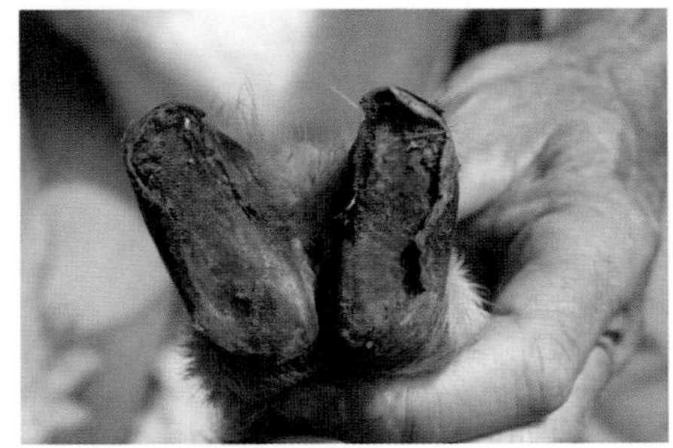

Sheep foot – no need to trim.

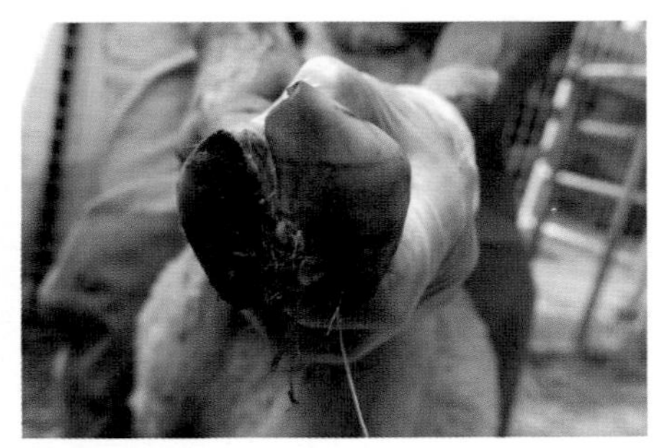

Overgrown sheep foot.

hooves. Lameness is often the first sign of ill health, and might indicate anything from a mechanical injury (such as bruising, a knock, treading on something sharp) to intestinal worms or a virus. Lameness can be serious, causing pain and affecting appetite and long-term health. Check all livestock daily to make sure nothing is hopping, walking unevenly, or eating on its knees, and take action if you find lameness.

If you have a lame bird, check for foot and leg problems first (such as heat in the leg, swelling, a wound, broken bones), and if you can't find anything mechanical, check its breastbone and body condition to see if it's underweight. If it is a mechanical injury, bed rest in isolation from other birds with close access to water and feed is indicated, with appropriate treatment for any wound. If there is heat in the leg that indicates infection, antibiotics may be required. Bumblefoot in birds is a hard swelling on the underneath of the foot that is normally caused by bacteria entering a small wound. Clean the site and remove any crust and pus, and treat with antibiotic spray.

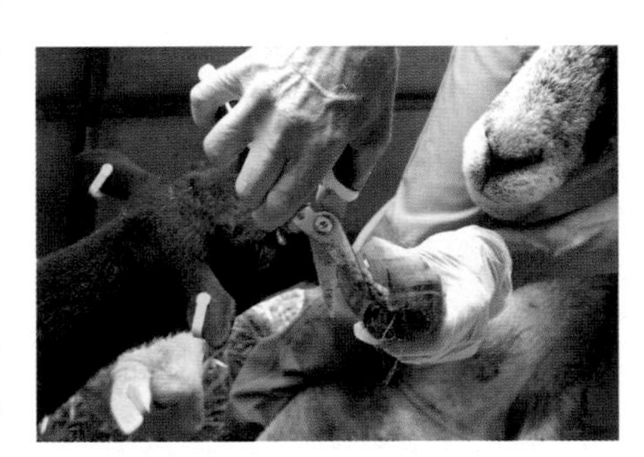

Trimming with foot shears.

We rarely see lameness in our cattle, and their foot care consists of having a professional hoof trimmer visit once a year to remove excess growth. Sheep and goats are a different matter, and their foot care is something that the smallholder is expected to manage. Current best practice confirms that trimming is both overused and counter-productive, causing more severe lameness and slower healing. Diagnosing the causes of lameness (with your vet if necessary) will enable you to treat accordingly; trim only if hooves are genuinely overgrown, and once any infections have been treated and inflammation is under control.

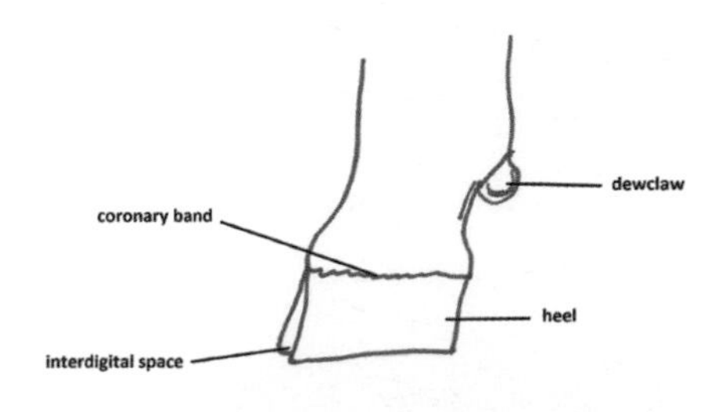

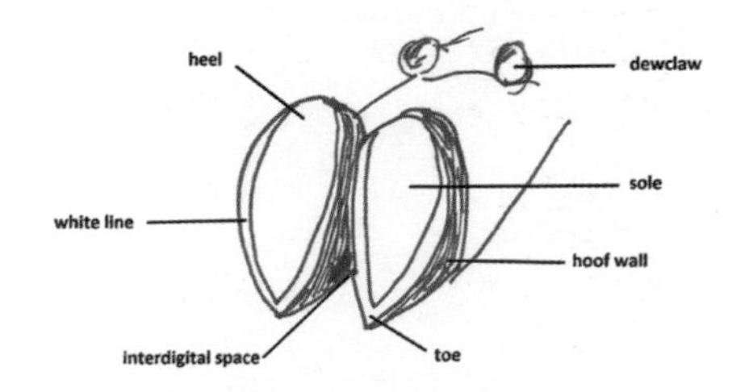

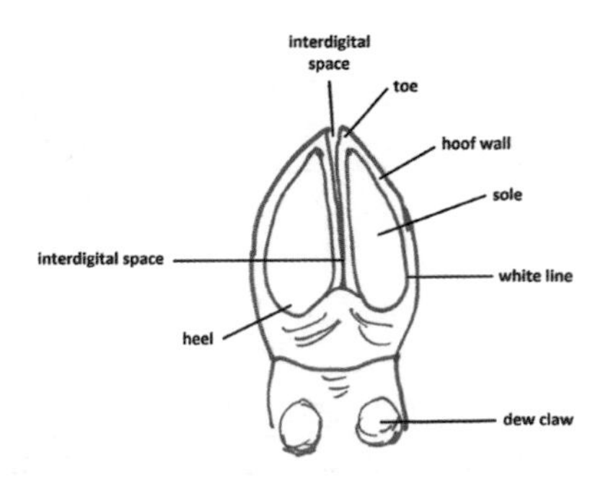

Goat foot anatomy (very similar for sheep and cattle).

Trimming a sheep's foot.

Overgrown cattle foot before trimming.

Cattle foot after trimming.

Goats tend to need their hooves trimming more regularly than sheep, but it's still important to appreciate when trimming is or is not appropriate. Goats and sheep that live on stony or hard ground may never need to have their feet trimmed, while those living on damp pastures may need attention every couple of months. Even vets have widely divergent opinions on the desirable frequency for hoof trimming goats, varying from every six weeks to once a year. The best approach for sheep and goats is to check for lameness daily, look more closely at their feet every month, and only trim when genuinely required.

When trimming, remove small amounts at a time to ensure you don't cause bleeding, and aim for the base of the foot to match the angle of the coronary band. Trim the feet of individuals that are sound first, and attend to any that are lame after that. Disinfect the foot trimmers if you've used them on a lame animal or one with any of the issues described below so that you are not transmitting infection from one animal to another.

Whereas sheep have their feet checked while sitting, goat's hooves are managed with the goat in a standing position, picking up each foot so that you can see the sole. If you have a milking stand this is ideal for checking and working on feet; if not, use a halter to tie up the goat while you attend to it.

Pigs mostly don't require their feet trimming; some adult individuals may have feet that grow overlong and need attention, but this is not likely for smallholders rearing weaners. The big issue is restraining large pigs: if you have a very relaxed pig you can trim their toes with horse hoof nippers while they are lying down; however, if that's not possible you may need to get the vet to sedate them (there are online videos showing people subduing their pigs for foot care by feeding them beer, but I can't say I've tried it).

Puncture Wounds

Hedge trimmings are a particular hazard for goats and sheep, and as time-consuming as this may be, do remove them from the pasture. Blackthorn and

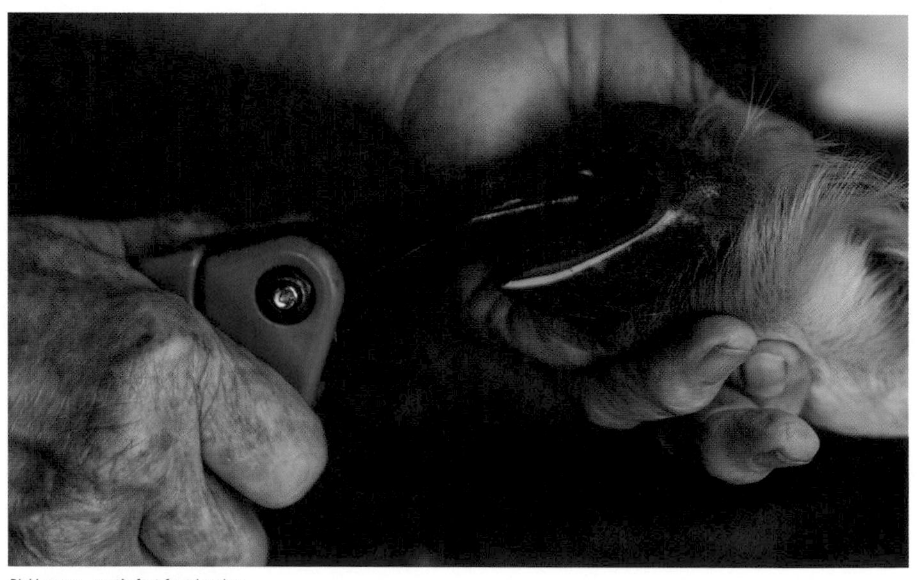
Picking up a goat's feet for trimming.

hawthorn are particularly pernicious, piercing feet and causing infection, and small stones and twigs can get caught between the claws. When putting up or repairing fencing be sure to remove all pieces of wire and stray fencing staples – exactly the sort of foreign bodies that a cloven hoof should not come into contact with. Noticed quickly, removing a foreign body from a foot can have instant rewards. Once removed, allow any pus to drain, clean the wound, and squirt with antibiotic or antibacterial spray.

Scald (Interdigital Dermatitis)

Scald is caused by bacteria, it is infectious, and is a common cause of lameness in kids and lambs in wet conditions in late spring. The skin between the claws becomes pinker and swollen, and may smell bad, with a thin layer of white discharge. Individual cases should be treated using antibiotic aerosol sprays. When several animals are affected, treating all goats or sheep in the group with a 10 per cent zinc sulphate solution or 3 per cent formalin in a footbath can provide effective control. After footbathing, sheep/goats must stand in a dry area (not in mud) so that the solution can dry on the feet. It may be necessary to repeat the foot bathing at weekly or fortnightly intervals. Foot trimming is not appropriate for scald.

Footrot

Footrot begins with scald and is extremely painful and contagious; sheep and goats with footrot are very lame. There is swelling and moistening of the skin between the claws, with infection spreading into the foot that separates the horn tissue from the sole; it can extend up the hoof wall when treatment is neglected. There is a foul-smelling discharge, and in severe cases the whole hoof horn might be shed.

Treatment for mild cases is by antibiotic aerosol spray, and where more serious, injection of long-acting antibiotics is also needed. Affected individuals should be isolated until the foot is healed to stop the spread of infection. Any loose horn should only be trimmed once the footrot lesion is healed. Footbathing is not an effective way to treat footrot as the substances will cause pain, and will increase the risk of granulomas forming.

White Line Disease

Penetration of the white line – where the wall and the sole join – by stones or sharp objects can cause lameness and infection, progressing to a white line abscess, which bursts out at the top of the foot. White line disease and abscesses require careful foot trimming to remove the foreign object and allow the pus to drain. Antibiotics will be necessary if the foot is hot or swollen.

Laminitis

Laminitis is a painful condition normally connected with ponies on too rich a diet, but it is also known in goats, sheep and cattle. It can appear after a female has given birth in association with a retained placenta, metritis (infection of the reproductive tract), mastitis and pneumonia, and in geriatric animals, or from acidosis after eating too much grain. Signs include unwillingness to move or stand, shifting from one foot to another, walking on the knees, teeth grinding (an indication of pain), with the feet tender and hot to the touch.

Anti-inflammatories will be needed and grain should be removed from the diet. Animals with laminitis may develop deformed, hardened hooves and require regular hoof trimming to deal with the overgrowth of the horn.

Contagious Ovine Digital Dermatitis (CODD)

Livestock with CODD are severely lame, typically from infection in one claw of a single foot. Ulcers develop at the coronary band (the junction of the hoof wall and the haired skin of the pastern), and can become severe. In the early stages not all affected animals will be lame, so careful examination of purchased animals for signs of CODD is essential. Treat by antibiotic injection and spray, and in all livestock avoid over trimming.

Toe Granulomas

Toe granulomas (also known as strawberry footrot) are painful red swellings of tissue caused by over-zealous trimming, untreated lesions, or chemical irritation from too frequent use of formalin footbaths. Treat with antibiotics if infected, and bandage the foot with copper sulphate; even with treatment they may not improve.

Shelly Hoof

Frequently found when animals are pastured on wet ground, the outer wall of the claw becomes loosened and forms a pocket, home for unwanted soil and debris. Carefully trim away loose horn and remove any impacted material. If infected, use antibiotic spray.

Old Age and Osteoarthritis

In pet herds elderly animals are quite common and they can become arthritic, just like humans. Anti-inflammatories can be given as pain relief.

MASTITIS

Mastitis is inflammation of the udder caused by bacteria, viruses or fungi, typically infecting through the teat or via damage to the teat or udder. Mastitis is a particular risk in (although in no way restricted to) dairy animals, is often contagious, and requires antibiotic treatment. The following are signs of mastitis:

- Swollen udder (half or whole).
- Udder or teat redness or misshapen udder.
- Hot udder.
- Pain when the udder is touched.
- Abnormal milk (clots, clumps, blood, watery colour), including change in milk taste or smell.
- Drop in milk yield.

- Lameness.
- Hungry young as the dam doesn't allow them to feed.

CLOSTRIDIAL DISEASES

Clostridial diseases (enterotoxaemia, pulpy kidney, dysentery, braxy, black disease, black leg, tetanus, botulism) can strike without warning; the organisms responsible are widespread in soil and invariably fatal, and all goats, sheep and cattle are at risk. Vaccination is required as a preventative, and is a routine task for smallholders. Vaccinations are administered subcutaneously in the neck.

Unless you can be certain that they have been vaccinated, it is advised to give bought-in livestock two doses, four to six weeks apart when they arrive. An annual booster dose is then given to ewes four to six weeks before their due lambing date to maintain immunity and provide high levels of antibodies that can be passed on to their lambs via their colostrum. When vaccinating in-lamb ewes do not forget to vaccinate rams and any other sheep on the holding. At a minimum age of three weeks lambs should receive the first of two injections, separated by an interval of four to six weeks.

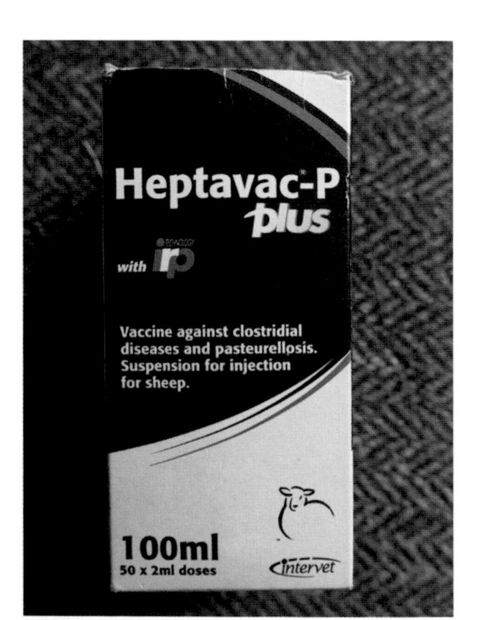

Vaccine for preventing clostridial diseases in sheep.

Vaccination of dirty or wet sheep should be delayed until the fleece is clean and dry, otherwise there is a risk of introducing bacteria with the vaccine and developing an abscess at the injection site.

Goats that have received their primary course need a booster every six months. Time these so that pregnant does receive a booster two to four weeks before kidding – this not only protects the doe, but passes protection to the unborn kids. Kids will need their primary vaccinations at ten to twelve weeks old, with a second dose four to six weeks later, and then continue with boosters every six months.

These vaccines have a shelf life of ten hours once the vial is broached, so a new bottle is required each time.

BLOAT

Bloat is the accumulation of gases in the rumen, which shows as distention on the left side of the abdomen and a loss of interest in food; it is usually caused by eating too much hard feed or grain, or grazing on lush pasture (frothy bloat), or growths in the oesophagus (obstructive bloat). Urgent attention is needed. Drenching with bloat drench or 100–200ml of vegetable oil and massaging the abdomen can help in cases of frothy bloat, and for obstructive bloat a stomach tube fed down to the stomach will release the gas, or the vet can make an opening into the rumen through the skin.

ZOONOSES

Zoonoses are diseases than can be transmitted between human and animal, so particular care needs to be taken.

Orf

Orf is a type of contagious dermatitis virus: blisters form around the nose and mouth (normally of lambs and kids), and the infection is transferred to the dams' teats through suckling. It enters through scratches and small wounds caused by browsing on thistles and the like. Although it usually clears up in four to six weeks, it can become so painful that the lamb won't feed. Use antibiotic spray to deal with any secondary infection where the skin is broken. A vaccine is available, but may not be appropriate in all situations, so discuss this with the vet.

When handling and treating orf, wear disposable gloves and wash your hands thoroughly afterwards as orf is painful and long lasting in humans. In people, orf starts as a small, red, itchy or painful lump, normally on the fingers, hands, forearms or face. Other possible symptoms include a high temperature and general tiredness. It usually clears up within three to six weeks, and cannot be spread between people.

ZINC SULPHATE CREAM

A big pot of zinc sulphate cream used for nappy rash is an item for everyone's livestock first-aid kit for treating minor wounds and sores. Note that zinc oxide is toxic to birds, dogs and cats if ingested in large quantities, but applying a thin layer to sore skin is seen as safe and beneficial.

Ringworm

Ringworm is a fungal infection causing circular crusty lesions around the head and neck of cows, sheep, goats and pigs. It can be transferred to humans by handling infected livestock, or any of the fixtures with which the animal has come into contact. When treating infected livestock, or disinfecting areas with which they have come into contact, wear gloves, and don't rub your face while doing so. Treatment of the animal with antifungal preparations may shorten the healing time, but normally the infection clears up in a couple of months.

Toxoplasmosis

Toxoplasmosis is a parasite carried by cats that is infectious to livestock and humans. If cat faeces contaminate feed and bedding, the parasite can cause damage and death to unborn lambs, kids, calves and piglets. Pregnant women should not be involved in calving, lambing, kidding or farrowing, or handling afterbirth or aborted foetuses, nor should they handle the clothes of anyone who is.

QUARANTINE

When bringing home additional livestock, a minimum twenty-one-day quarantine (and twenty-eight days is better) is recommended, during which new livestock are kept completely separate from other animals. Regarding goats, if you have concerns about CLA, at least two months quarantine is needed for monitoring signs of developing abscesses and swollen glands.

The correct quarantine treatments should be used for incoming livestock to remove resistant worms. Dosing animals and then putting them straight on to clean pasture increases the risk of wormer resistance developing on your holding. Ideally, contain new four-legged livestock on hard standing (or a small grassed area) for their first two days so that worms voided can't contaminate your pasture, and put new poultry in separate areas away from your existing flock.

Once wormed, take a dung sample seven to fourteen days later depending on the wormer used, and have it checked to ensure the wormer has been successful; if not, worm again with a different family of wormer. Do visual checks for lice in multiple places over the incoming animals, and also for footrot and CODD.

MEDICINE DATASHEETS

Datasheets for all veterinary medicines (for when you lose the box or paper insert that tells you withdrawal periods, dosage, route for administration etc) are all available online at www.noahcompendium.co.uk.

MOULTING

Moulting is an entirely natural process; ducks moult twice a year, geese and chickens once a year, from around eighteen months old in the autumn, although some like to start in summer. It takes two to three months for new feathers to grow in fully. Cows will shed their winter fuzziness as spring approaches, and by summer have a smoother coat, and pigs will also shed their bristles; these regrow as the weather cools. Self-shedding sheep breeds such as the Wiltshire Horn will drop their fleece as the weather heats up.

PROBLEMS IN POULTRY AND WATERFOWL

Breathing Problems

In poultry and waterfowl, gasping for breath can indicate aspergillosis, which comes from fungal spores that develop in mouldy bedding or food; it can therefore be avoided by good management. Ensure bedding is always dry and changed regularly, that feeders are kept clean, and mouldy food is disposed of where birds can't get at it.

Gape worm also causes breathing problems: treat with wormer and move birds to clean ground, add lime to areas that have been heavily used, and leave the area free of birds for a period. An obstruction may also cause breathing problems in poultry and waterfowl, so keep their areas free of non-food items they might swallow.

Mycoplasma shows as a discharge from the nostrils, sneezing and bubbling in the corner of the eye, and an unpleasant smell emanating from the bird's nostrils. Antibiotic treatment, given by injection or added to the drinking water, will be required.

Egg Issues

Young layers may have difficulty laying eggs, and a youngster looking hunched may be egg bound. Warm (not hot) vegetable oil rubbed around and just inside the cloaca (vent) may help, as does putting the bird in a dark space under a heat lamp. Gentle manipulation may help, but avoid breaking the egg inside the cloaca as damage from a broken shell can lead to blood poisoning. If shells are thin or soft, address this through improving the diet, making poultry grit available, and offering calcified seaweed.

Wing Problems

Occasionally birds can have minor wing deformities, which are unsightly but don't seem to impede the bird; however, don't use these birds for breeding. Other problems, such as angel wing where a single or both wings hang down, are caused by too much protein in the feed. Ensure young birds are on the correct diet, moving them on to a grower ration as soon as their feathers start to appear. If one or both wings hang down, strap them closely to the body for a week or ten days (micropore tape is useful for this), replacing the tape every couple of days.

HUMANE DISPATCH

If livestock is suffering and beyond treatment, they should be euthanised humanely. Your vet can do this for you, or the knacker, or you can follow the guidance from the Humane Slaughter Association.

SAFE DISPOSAL OF SHARPS AND UNUSED MEDICATION

Use a proper sharps bin to dispose of used hypodermic needles, syringes with needles, scalpels and so on; this can be bought from your vet to include the cost of disposal when full. For any unused/out-of-date medication a 'doop' ('destruction of old pharmaceuticals') bin bought from your vet will also include the cost of disposal when full.

Doop and sharps bins.

TOXIC PLANTS

A number of plants are very toxic to livestock, while some food is fine to feed in small quantities, but should not be given in large amounts. Brassicas contain a compound that blocks iodine absorption, and as iodine is essential for the development of a foetus, brassicas should not be fed to pregnant dams. Plants are toxic in a number of ways, from causing sudden death (yew, rhododendron, foxglove, laurel, water dropwort, linseed), to those that cause scouring, haemorrhage, photosensitisation, bloat, liver damage, anaemia, stillbirth, vomiting and so on.

Some plants, such as buttercup and sorrel that grow naturally in pastures, shouldn't cause issues (ingesting large amounts causes kidney damage) as long as there is plenty of alternative food available, as they taste bitter to herbivores. Some plant parts are toxic when others are not; ivy leaves are fine for livestock to eat, but ivy berries are not. When pruning ornamental shrubs keep all their trimmings away from livestock.

Signs that livestock has ingested something poisonous includes salivation, lethargy, depression, vomiting, scours, muscle tremors, and a bloated/painful abdomen. Talk to your vet about any emergency medication you should have on hand if plant poisoning is suspected, as treatment may need to be given by the keeper swiftly before a vet can be present. They may suggest activated charcoal, vegetable or mineral oil, Milk of Magnesia, or prescription-only medications.

Plants Toxic to Livestock

Aconite
Anemone
Alder
Avocado
Azalea
Black nightshade
Box
Bracken
Brassicas
Broom
Bryony
Buckthorn
Buckwheat
Buttercup
Charlock (wild mustard)
Clover
Common sorrel
Daffodil
Deadly nightshade
Fool's parsley
Foxglove
Gladiolus
Ground elder
Hellebore
Hemlock
Honeysuckle
Hydrangeas
Ivy (berries only)
Laburnum
Laurel
Lesser celandine
Lilac
Linseed
Lobelia
Lucerne
Mayweed
Meadow saffron
Mullein
Oak (acorns)
Old man's beard
Potato leaves
Pieris (forest flame)
Pine
Privet
Prunus family (apricots, almonds, cherries, nectarines, peaches, plums, sloes)
Ragwort
Rape
Rhododendron
Rhubarb
Spindle tree
St John's Wort
Tomato (leaves only)
Tormentil
Tulip
Virginia creeper
Walnut
Water dropwort
Woody nightshade
Yew

14 Meat from the Smallholding

THE REALITY OF REARING FOR MEAT

I started writing this chapter the day a favourite sow went to the abattoir. She had been with us for many years, was a joy to handle and be around, and produced lovely litters – but now it was no longer appropriate to breed from her. We gave her a care-free eight-month holiday, asked people who had previously wanted to home her in retirement to take her on but were told their circumstances had altered, and finally knew she had to be culled. Land isn't elastic and there were other pigs needing the space. Our choice was to have her euthanised, or turn her into sausages.

Having a big demand for mutton from older ewes, and mature beef from cows that had served their time as fruitful breeders, taking older animals that had been stalwarts and old friends to the butcher was nothing new. But this didn't stop me thinking about my decision every time I fed her, or heard her grunt companionably in her paddock. Others might have taken her off to market, unable to take her on her final journey – but that would have been both irresponsible and unforgivable. It is our responsibility to make sure that our livestock do not have a miserable time being shunted hither and yon, travelling who knows how far in the company of strangers, to their ultimate inevitable destination. Taking our old sow directly to slaughter meant we had control of her care until the end.

You can have great meat from native livestock that has been reared well, producing wonderfully sustaining protein and fat. But the reality is that in order to eat meat, something has to be killed, and this is the challenge that faces all smallholders who hope to produce their own meat. So how do you deal with taking to slaughter an animal to whom you have become attached?

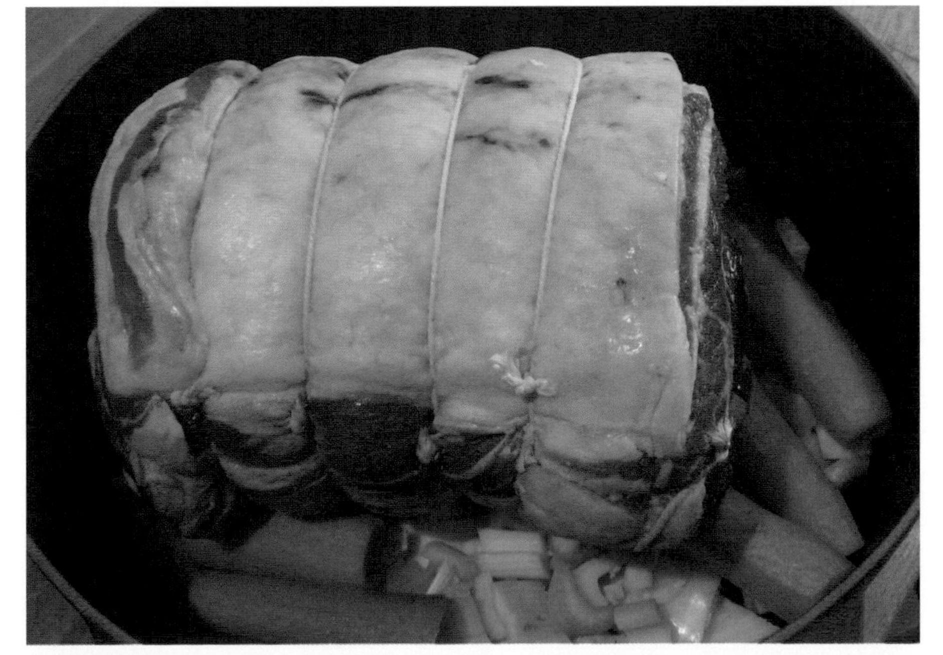

Devon Ruby brisket.

I can only share my own perspective and experience, which is that you give your animals the very best life you can, you are forever aware that you are rearing them for meat, and you do your utmost to provide a quick, painless and stress-free end. In the lead-up to butchery time channel any concerns you might feel into ensuring that all the practicalities have been dealt with: planning how you will load and transport your animals calmly, completing the necessary paperwork, checking for any road closures or diversions that may affect your journey, and so on.

Producing your own meat is very rewarding; it has minimal food miles, and you have complete control over how the animal is reared, fed and treated. Your meat might be intended entirely for home consumption, or for sale, and in both instances you'll want to produce the very best possible. Certainly it's a primeval moment when you look at a chicken and contemplate what it will look like oven ready, but when you start to pick up your birds not for a cuddle but to test the quantity of breast meat under all that feather, you are mentally prepared for feeding yourself on home-reared meat.

Breeding with Intention

Breeding should always be carried out deliberately, with forethought and intention. Casually bringing livestock into the world without consideration for how you will care for and manage them is irresponsible, and is why I get somewhat hot under the collar when schools hatch eggs in the classroom (which can be a wonderful thing), and have no plan for what they will do with the resulting birds. Breeding because young animals are cute (they are) and might be fun (they will be) rather than because they will have a purpose, or by mistake because you think an ageing female won't become pregnant when kept with an adolescent male of the species (she will), is storing up all sorts of problems for the smallholder.

Quail, dressed and ready for cooking.

BEING MOTHERLY

A word of caution about breeding livestock being motherly: it's a word that conceals hidden dangers for the unwary. Motherliness is a positive-sounding trait, depicting a dam that cares for her young, and it is critical for the survival of offspring, and that care includes protecting newborns with vigour. Being pecked by a hen as you check her chicks is one thing – you may sustain a mildly nipped or bruised finger. But having a goose savage you is less comfortable, and a cow tossing you against a wall as her hormones rage after she has calved could be downright lethal. Therefore never be blasé about your positive bond with your livestock, particularly when they have newborns.

Livestock being reared for meat needs to be from suitable breeds. There's no point raising a marvellous flock of runner ducks for meat when what they excel at is egg laying; their carcase may be delicious, but the quantity of meat is scanty. A bantam hen will give you a poussin-sized bird, while an Orpington of whatever colour delivers 2.5kg or more of superb chicken. That's not to say that small can't be beautiful: from quail to small-breed sheep, their meat may not be plentiful but it can still be outstanding in quality.

Do you have to breed your own livestock to produce meat? The simple answer is no, you don't. There are many ways to buy in animals and take them to meat weight, rearing them for an appropriate period depending on how old they are when you bring them home.

BODY CONDITION SCORING

Weighing and visual checks are a starting point for checking body condition (too thin, too fat, just right), but for accurate assessment, handling the animal is essential. Too thin and too fat animals have poorer conception rates, increased risk of birthing difficulties, and are more prone to other health problems. Body condition scoring (BCS) is a hands-on method for assessing the amount of fat and meat carried by an animal.

Typically, livestock is scored before mating to ensure males and females are fit for breeding; to check females a couple of months before they give birth to manage their diet; to check developing young to ensure they are putting on weight and condition; and when choosing animals to take to slaughter. Two things govern the quantity of meat on a carcase: conformation, which is largely hereditary; and body condition score (BCS), which is the amount of meat and fat cover within the limitations of the breed.

For BCS, a five-point scale is used for all species, where a score of one is assigned to a very thin animal, and five to a very fat one. Scores are measured slightly differently for sheep, goats, cattle and pigs.

Goats

Feel for fat cover with your hands over two areas of the animal's body: the loin and rib (spinous and transverse processes), and the breast (sternum). The sternum is the bone running along under the goat, at the brisket between the two front legs. Healthy goats should have a BCS between 2.5 to 4, whereas those with a BCS of 1 to 2 indicate a management or health problem. At mating and at kidding, does should have a BCS of 3 to 3.5, and by weaning should be no less than 2.0. At slaughter, goats should have a BCS of 2.5 to 3.

Sheep

Feel for fat cover with your hands over the top of the tail where it meets the body (dock), the loin and shoulder, the rib and the breast. At tupping, hill breeds of sheep should have a BCS of 2.5, upland ewe breeds a BCS of 3, and lowland breed ewes 3.5. They should lose no more than 0.5 BCS throughout pregnancy and lambing. Rams should be 3.5 to 4 before tupping. At slaughter, sheep should have a BCS of 3.

Cattle

Handle animals on their left side as seen from behind; a full rumen on the right can make the cow appear fatter than it is. Feel for fat cover around the tail-head and pelvic (hip and pin) bones, the ribs and the loin. You are looking for a BCS of 2.5 to

Body Condition Scoring

Body condition score for all species	Sheep, cattle and goats	Goats only
BCS1		
Much too thin. Looks emaciated, backbone visible with continuous ridge, flank hollow, ribs visible, no fat cover. Spinous and transverse processes are prominent and sharp, and can be grasped easily in the fingers. Loin is thin with no fat cover. **Cattle:** Tail-head has deep cavity with no fat under the skin. **Goats:** Sternal fat is easily grasped and moved from side to side, joints joining ribs and sternum are easily felt.	*Loin and rib score 1.*	*Sternum score 1.*
BCS2		
A little too thin. Thin, backbone visible with continuous ridge, some ribs can be seen, little fat cover. The spinous processes are prominent but smooth, individual processes felt only as corrugations. Transverse processes smooth and rounded, but possible to press fingers under. Loin muscle present with little fat cover. **Cattle:** Tail-head shallow cavity, some fat under the skin. **Goats:** Sternal fat is wider and thicker but can still be grasped and lifted. Joints are less evident.	*Loin and rib score 2.*	*Sternum score 2.*
BCS3		
In good condition. Backbone is not prominent, ribs are barely discernible with even fat cover. Spinous processes smooth and rounded, and bone only felt with pressure. Transverse processes also smooth and well covered, and hard pressure needed to find the ends. Loin muscle is full and with moderate fat cover. **Cattle:** Fat cover over tail-head. **Goats:** Sternal fat wide and thick with very little movement when grasped. Joints are barely felt.	*Loin and rib score 3.*	*Sternum score 3.*
BCS4		
In fat condition. Backbone and ribs can't be seen, sleek appearance. Spinous processes only detectable as a line. The ends of the transverse processes cannot be felt. Loin muscles full and rounded with a thick cover of fat. **Cattle:** Tail-head filled with a fold of fat. **Goats:** Sternal fat difficult to grasp because of width and depth, and cannot be moved from side to side.	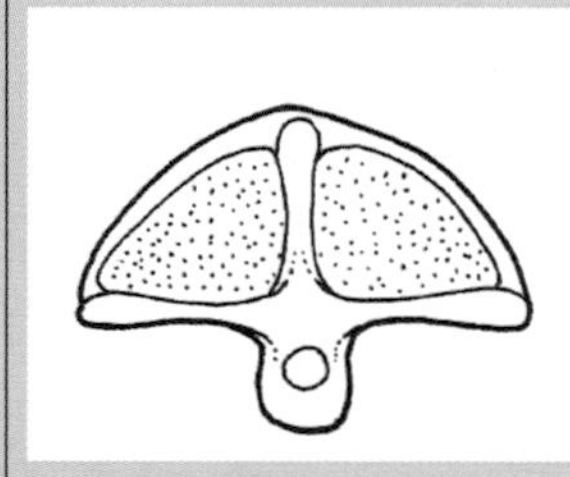 *Loin and rib score 4.*	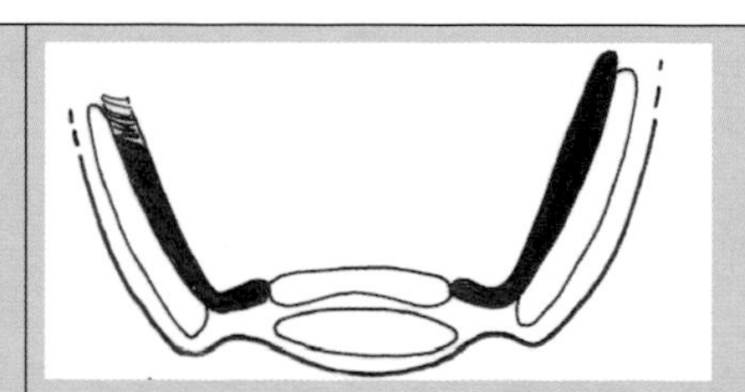 *Sternum score 4.*
BCS5		
Much too fat. Backbone buried in fat, ribs not visible and covered with excessive fat. Spinous and transverse processes cannot be detected with pressure, and dimpling in fat layers. Loin muscles are full and covered with thick fat. **Cattle:** Tail-head almost buried in fat. **Goats:** Sternal fat extends and covers the sternum joining the fat over the ribs and cannot be grasped.	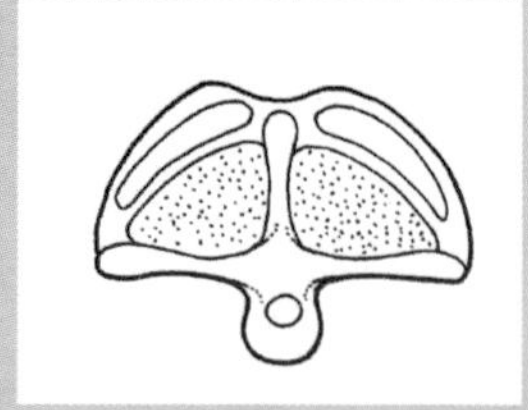 *Loin and rib score 5.*	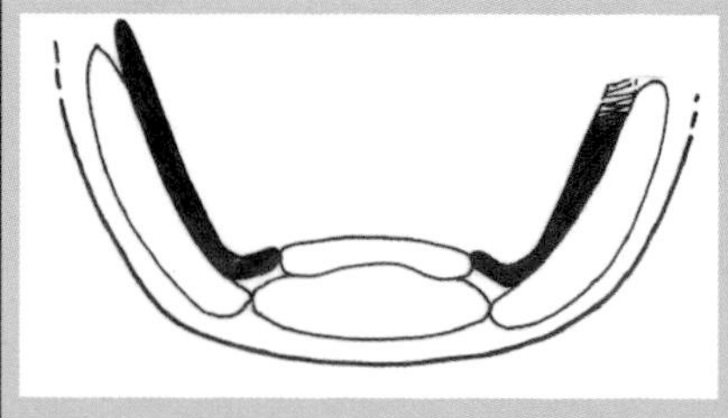 *Sternum score 5.*

3 at calving and for most of the year. If you house your cows over winter they are likely to lose a bit of condition, so aim for a BCS of 3 to 3.5 at housing. At slaughter, cattle should have a BCS of 3.

Pigs

Feel for fat cover with your hands at the shoulders, backbone, hips and tailhead. Sows should farrow with a body condition score of 3 to 3.5, and at weaning they shouldn't be less than 2.5. At slaughter, pigs should have a BCS of around 3.

For all livestock, males grow more quickly than females and tend to be leaner. This has a measurable impact when rearing beef, as they have a longer growing period than lambs and weaners. We take steers to thirty months or more, and heifers reared for meat tend to be ready at twenty-four months; any older means the females may be carrying more fat than is desirable.

KILLING OUT PERCENTAGES (KO%)

Killing out percentage describes how much finished carcase weight (that is with the head, feet, skin and guts removed) is obtained from the live animal, and is expressed as the deadweight of the carcase (not the fully butchered weight) in relation to the weight of the live animal immediately before slaughter. The final butchered weight will be lower than the deadweight, and the more the butcher does (boning, rolling, trimming, dicing) the less the resulting weight will be.

Lambs

Commercial lambs are slaughtered at around 42kg liveweight, resulting in a 19kg carcase, giving a KO percentage of 45 per cent (19 divided by 42 multiplied by 100 = 45.2 per cent). If you have rare, hill or primitive breeds, your KO percentage is likely to be somewhat lower, between 35 and 45 per cent, and the weights will be lower too. There is also a difference between the sexes: entire ram lambs (and wethers) have a lower KO percentage than a ewe lamb. Lambs with heavy fleeces or horns will also have a lower KO percentage than those with shorter wool, or which are polled.

Beef carcases in the butcher's chill room.

Goats

The deadweight carcase of six- to twelve-month-old meat goats raised on pasture will weigh in the region of 14–20kg (32–40kg liveweight) or more, and dairy-cross males reared specifically for the restaurant market fed ad lib until they reach 55–60kg liveweight can achieve as much as 24kg deadweight. At twelve months you should expect a killing out percentage of around 45 per cent, so a 40kg kid should deliver an 18kg carcase.

Pigs

Once fully butchered, anticipate approximately 60 per cent of liveweight plus the offal, head and trotters, which rounds up the yield to 65 per cent.

Cattle

Expect approximately 60 per cent of liveweight as your deadweight before butchery, and 35 per cent of liveweight as finished butchered weight.

Poultry

The butchered weight for chickens, turkeys, ducks and geese is approximately 70 per cent of liveweight once neck and giblets are included.

HOME SLAUGHTER

There are two lawful ways in which to have your animals slaughtered and prepared for your own consumption:

- In an approved slaughterhouse.
- On your farm by you or a licensed slaughterer (or a holder of a Certificate of Competence) under your responsibility and supervision. It is

Meat rabbits.

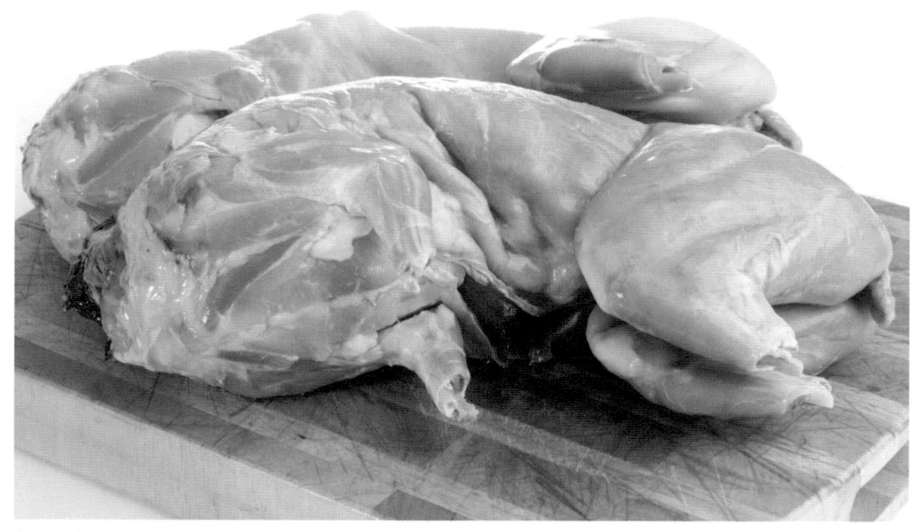

Oven-ready rabbit.

unlawful in all cases to employ anyone apart from a licensed slaughterer to slaughter animals on your property.

Carcases or parts of carcases from home slaughter livestock, other than those being disposed of as animal by-products, cannot be removed to another premises for processing.

For the vast majority of smallholders the slaughter of livestock is best left to the abattoir. Home slaughter is not appropriate for anyone who doesn't have the skill to dispatch an animal without causing it any suffering. However, there are sound arguments if you *do* have the skill (and not misplaced confidence) for slaughtering at home, particularly if your nearest abattoir is far away and you want to eliminate the stress of a long journey on livestock.

You don't need a licence to kill your own pigs, sheep, goats, cattle, poultry, rabbits and hares to eat at home, provided that you own the animal, kill it on your property, and are killing it for you or your immediate family to consume (all of whom must live on your property). The sale of poultry slaughtered on your holding is allowed, but anyone carrying out slaughter of poultry for sale/use beyond the immediate family must hold a Certificate of Competence. For information on poultry slaughter training, contact the Humane Slaughter Association.

For all species being killed at home you need to know how to restrain, stun and kill the animal humanely and quickly, and avoid causing it any avoidable pain, distress or suffering – so familiarise yourself with the guidance from the Humane Slaughter Association. An animal must always be stunned before killing, making it immediately unconscious, and it must remain unconscious until it is dead, killed using a method such as bleeding, immediately after stunning.

There are strict regulations around the disposal of animal waste, which includes skins not being tanned, blood, skull, brains, spinal cord, eyes, spleen and any intestines you don't intend to consume; these must be stained blue and disposed of according to Transmissible Spongiform Encephalopathies (TSE) regulations as Specified Risk Material. This disposal is beyond the capabilities of most smallholders.

ABATTOIR BOOKING AND PROCEDURE

Taking those first animals off to slaughter is the moment that most new livestock keepers fear. The week before the circled date in the diary you inevitably remember how cute they were when you brought them home or helped them being born, and grow pale at the prospect of turning them into meat. You sternly remind yourself that the main reason for doing this is to provide fabulous food, that you are fed up with horsemeat scandals, dubious welfare practices and ruminants stuffed with soybean meal rather than natural forage. It's entirely appropriate to be nervous, and it is therefore crucial to know what is required so that you focus on what needs to be done in a manner that does not stress your animals on their final journey.

First you need to find an abattoir. Local farmers, smallholders and butchers will be able to recommend abattoirs. Very few abattoirs have their own website, but the Food Standards Agency website lists approved butchers, slaughterhouses, cutting and processing plants.

If you are having the meat back from the abattoir for your own consumption or to sell on, this is known as a private kill: the meat is not being sold to the abattoir. Not all abattoirs take every kind of livestock: some specialise in pigs, others in cattle, a few do nothing but poultry (although there aren't many poultry abattoirs that do private kills), those that take sheep will also take goats, and some will do all of the four-legged species. Abattoirs don't necessarily kill every day, and those that take more than one species usually have specific days for each species.

When taking livestock to the abattoir you need to comply with three legal requirements as described in Chapter 4: movement licences (passports for cattle), medicine withdrawal periods and identification (such as ear tags).

If you wish to have skins or horns back from the abattoir you need to register with the Animal and Plant Health Agency (APHA), completing and submitting the relevant form, downloadable online; permission is usually granted within a few days of submission.

Transport

You may have brought your weaners, lambs or goatlings home in the back of a 4×4, but you can't transport the finished animal to the abattoir in the same way: you'll need a fit-for-purpose livestock trailer. Although you may be stressed, most animals are easy to load into a trailer, particularly sheep and goats. Pigs need more encouragement, and I strongly recommend feeding the pigs in the trailer for a few days before slaughter day so that they trot in and out casually and without concern.

For cattle that are not halter trained, we find it easiest to back the trailer up to the cattle crush, run them through the race and crush, and into the trailer. Some straw scattered on the ramp is helpful, and make sure the ramp doesn't bounce up and down as they walk on to it. Clean straw should be provided in the trailer – a generous scattering rather than a deep mattress – but sawdust should be avoided as it adheres to the animal's coat and may cause problems when the carcase is dressed.

The Layout of the Abattoir

Abattoir layouts vary, in particular in the amount of room you have available for reversing your trailer to the lairage, the area where animals are kept in the short term while they await slaughter. Practise and perfect your trailer reversing skills at home. The animals are your responsibility while in the trailer; once off the trailer ramp they become the responsibility of the abattoir.

Don't expect help unloading (although some places will help, others may not; we find cattle abattoirs particularly helpful), so take a companion with you if necessary. Most animals trot out of the trailer with a bit of gentle shooing if you get in behind them. With pigs that may well be waking up from a sleep, or don't like the unfamiliar space, take a small bucket of feed to shake in front of them – this normally gets them trotting calmly out of the trailer.

YOUR TRIP TO THE ABATTOIR

You need to think about and organise the following when planning your trip to the abattoir:

- Check that the abattoir takes the appropriate species.
- If you are using a smaller abattoir you will need to book well in advance – sometimes several weeks – especially in the run-up to Easter and Christmas. Agree both the date and time of drop-off with the abattoir.
- You need a livestock trailer to get your animals to the abattoir (borrow or hire if necessary).
- Appropriate ear tags need to be in place before you leave your premises.
- Make sure any medication given has passed the withdrawal period so that the meat is fit to enter the food chain.
- Livestock must be clean (not wet, dirty or muddy), so keep them inside on a clean bed of straw for a day or two beforehand. A clean belly is critical to avoid any bacteria from the coat contaminating the meat, so clip the belly area of any long-haired goats or sheep (a 5–8cm-wide strip from brisket to tail).
- Livestock should be empty (this means no food for eight to twelve hours before you leave your holding) to minimise the risk of faeces in the intestine contaminating the meat. Provide water until they are loaded on the trailer.
- If the abattoir offers a butchery service, have ready a clear cutting list describing how you want an animal butchered. Ask the abattoir in advance for their cutting list options – most provide simple guidelines.
- If the abattoir doesn't offer a butchery service and the meat is for your own use, you can cut the meat yourself, and just have the animal slaughtered and the carcase split. If you intend to sell the meat and/or need it butchered, make arrangements for this *before* taking the animals to the abattoir. Either contact your local butcher to check if they will do private butchery and arrange with them which abattoir to use, or ask your abattoir which butchers they deal with and contact them to arrange collection and butchery of the carcase(s).

- Have your movement licence/passport completed, and if you do it online remember to print off and take the haulier form with you. You are responsible for producing the licence.
- You can clean out the trailer at the abattoir or at home (if the latter, you need to complete a form that commits you to cleaning it within twenty-four hours, or before the vehicle is next used for carrying livestock, whichever is the soonest; the abattoir will provide this form).
- Do you want the skins or horns? If so, you need to tell them. Pick them up within twenty-four hours, and ask for the skins to be salted if possible. When you get them home add more salt (on the flesh side), and arrange to get them to a tannery. If well salted they will keep almost permanently before taking them for tanning (*see* Chapter 17).
- If you want the offal, pick it up within twenty-four hours.
- Pick up the meat on the agreed date (arrange this when booking the slaughter date).
- You may prefer small on-farm abattoirs that are used to smallholders with a few animals, but large-scale abattoirs are very efficient and need less notice when booking.

Briefing the Butcher

If the abattoir offers a butchery service it's likely that they will offer a simple, standard cut; for example for sheep and goats the legs could be whole or halved, the shoulders whole or halved, the breast whole or boned and rolled, and the loin cut into chops or left as joints. If you want the offal, let them know or it will disappear. A local butcher is likely to offer additional cutting options, but be aware that the more complex your requirements, the more time it will take the butcher and the more they will charge.

If the meat is for your own use, think about how you like to cook and eat it, and create your cutting list based on this. You might always cook whole legs and shoulders for Sunday family gatherings, relishing any leftovers to make all sorts of supplementary meals. Or you might prefer small joints that will be demolished at one sitting. Your butcher can also bone and roll, dice, mince the breast, and make burgers and sausages.

If you are selling your meat they can bag or vacuum pack and label your cuts, giving you the whole lot back in a bag or cardboard box, or laid out on disposable trays, or in a courier-friendly box ready to send out to your customers. Suitable packaging and labelling that meets Food Standards Agency requirements is essential if you are selling your meat, and options that exclude plastic and are more environmentally friendly are being developed all the time. And note that you do not get four legs from a sheep or goat – their carcase has two legs (the hind legs) and two shoulders (the front legs).

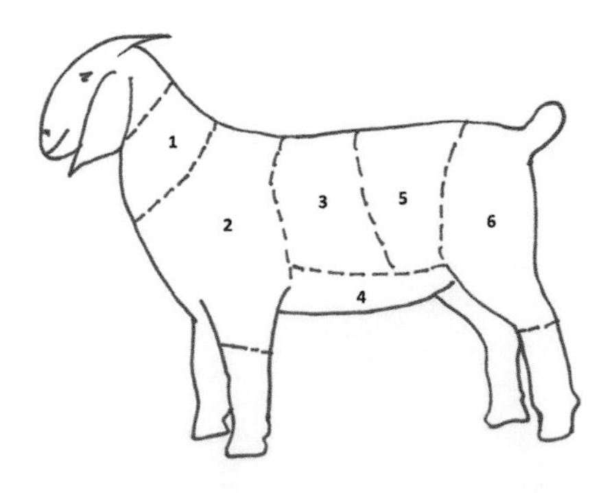

Sheep and goat meat cuts. Key: 1. Neck, 2. Shoulder, 3. Loin, 4. Breast, 5. Chump, 6. Leg.

Example Cutting List

	Cutting and other requirements	Notes
Number of lambs	Two	Date to abattoir Date for collection
Skins	Skins to be salted	Date for collection
Offal	Livers and hearts only (lights not required)	Date for collection
Whole lamb	Legs whole Shoulders halved Breast boned and rolled Loin as chops Chump steaks Neck Kidneys	Vac packed and labelled with weights. Chops in bags of two
Half lamb	Leg halved Shoulder whole boned and rolled Breast minced Loin as a joint Chump joint Neck Kidney	As above
Half lamb	Leg into joint and separate shank Shoulder cubed Breast whole on the bone Loin as chops Chump as steaks Neck	As above

Traceability

Smallholders always worry that they may not get back their own carcases from the abattoir or butcher; although I have heard of a few instances of this it is very rare, as carcases are tagged to ensure traceability. However, if you want the offal back – kidneys aside as they remain attached to the carcase – there's every chance that the liver and heart may not be yours. If you go to a small abattoir and want to be sure of getting your own offal, you'll probably have to wait until the animals are slaughtered.

Abattoir tag for lamb.

HOME BUTCHERY

There is nothing stopping you taking home the carcase once it's been chilled down (allow twenty-four hours) to do the butchery at home, if it's for your own consumption. There's no point, for example, in having a pig butchered into chops and joints if you intend to turn it into sausages at home; in this case fetch the carcase, debone it, mince the meat and an appropriate amount of fat, and proceed from there. There are some great videos online on basic home butchery for those who find demonstrations more helpful than books. Below you'll find guidance for dressing your own poultry.

Maybe your intention is to make the most of every bit of the carcase – skins, offal, horn. However, it's impossible, unless you work in an abattoir or do home slaughter, to get the blood back from your pigs, which means that making black pudding from fresh blood is a sad omission in the smallholder's drive for head-to-tail eating. Nor have I been able to get the tongues back from lambs that have gone for slaughter for many years. But you shouldn't have much trouble receiving pigs' heads and trotters, ox tongues and tails, every type of heart, liver and kidney, and even the lights (lungs), which are a core ingredient in haggis.

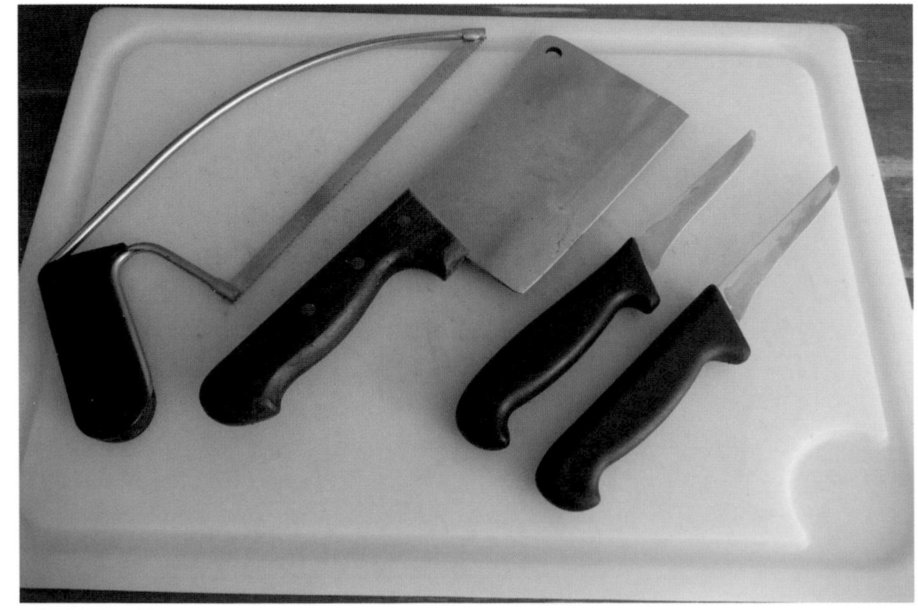

Butchery knives and saw.

REARING FOR MEAT

Rearing Goats For Meat

The majority of goats are dairy breeds, so is there anything special that you need to do for goats being raised for meat? Like all goats they need a balanced and varied diet to keep them fit and healthy, preferably a sustainable, natural and, as it happens, low-cost mix of browse, pasture and hay. A diet that limits or avoids concentrated grain-based feeds will result in a slower growing animal, but as for all ruminants, the resulting meat's taste and texture will be far superior, so it's definitely worth the wait. Good quality, sweet-smelling hay should be available for your growing goats at all times, and during the summer months a meat goat will graze well.

A goat being raised for meat should be meaty. Choose goats that are well muscled and have plenty of width and depth to their body, preferably with a good length of loin and a rounded rump. A chunky conformation rather than a scrawny outline is what you're after.

Meat goat kids are at their peak as regards taste at a year old, so there is no need to rush to finish them early; goat-meat producers in the UK take meat goats to slaughter from six to eighteen months. Different cultures have their own age and gender preferences for goat meat, ranging from milk-fed kids as young as two to three months to mature bucks.

Lamb and Hogget for Meat

Buying in bottle-fed 'orphan' lambs for meat requires a strong stomach and a stiff upper lip, as these lambs will treat you like their mother, running up to be fed, standing on your feet, nudging the backs of your knees and generally acting like your new best friend. Taking such affectionate creatures off to slaughter can test your resolve, so be warned. You may find it far easier to breed your own lambs for meat, who will retain their independence and focus on their natural mum rather than you. The majority of sheep-keeping smallholders choose to breed sheep, which means getting to grips with the demands of lambing and year-round husbandry; but buying in a group of orphan lambs can work very well when you have limited grazing.

Hogget is lamb in its second year. It's as tender as lamb, but has a deeper flavour (not a strong flavour – don't be put off), and is leaner as it has lived off its fat reserves over winter. The carcase will be slightly bigger too. Normally spring-born lamb would be taken off for meat around September, and as hogget the following May or later, once it's had a chance to put on some weight and flavour from the new spring grass – and I can't recommend it highly enough.

Meat from Older Animals (Culls)

Every livestock keeper needs to think about any animals that are no longer fit for purpose. You may have room for a pet or two, but you also need to think about avoiding over-stocking your land and over-taxing your time, energy and wallet with non-productive animals. If you are breeding you are not going to be able to keep everything unless your land and facilities increase, and at some point old breeding stock needs replacing with youngsters.

Having a clear rationale as to culling can be very difficult for new livestock keepers, who understandably want to keep all their old pals; with small livestock numbers you become hugely attached. But if you want livestock to be productive and healthy, you need to use your head as well as your heart. A cull policy may include dispatching barren females or ones that fail to deliver or rear live young, those with persistent foot or other health problems or who have prolapsed during birthing, females with inadequate milk, and those with traits you don't want to reproduce, or which exhibit a seriously challenging temperament.

All these animals can be used for meat, using the longer, slower cooking methods to achieve tenderness; they will have plenty of flavour. Arrange for the carcase to be hung for longer than would be the case for youngstock (a fortnight for old sheep and goats, six weeks or so for old cattle), and be sure to mark the returned joints as mature meat/mutton before you put it in the freezer, as failing to do so can result in cooking confusion and chewy disasters rather than wonderful meals later on.

We cull old cockerels after three or more years of service, always marking them as 'old cock' in the freezer. This ensures we casserole them for hours as a gorgeous coq au vin, rather than roast by mistake; the skin has lots of flavour but is best peeled off before eating unless you enjoy the rubbery knicker-elastic texture – but the meat itself is glorious.

SELLING YOUR MEAT

Regulations

There are regulations that you need to abide by when selling food. Contact the local Environmental Health Office as you are almost certainly going to have to register as a food business, and make that contact well before contemplating selling meat to anyone other than local friends and family. An Environmental Health Officer (EHO) will visit you to discuss what your intentions are, and to inspect the premises as appropriate to your proposed venture. You will have to meet all their requirements before advertising and selling your meat.

There is confusion over what sort of enterprises need to register as a food business. If you sell small quantities (as defined by your local EHO: it's not for you to define what is a small quantity, so you must ask them) of primary products – which includes simple cuts of meat but nothing processed – direct to your customer face to face, or to a local retailer who directly supplies the final consumer, then you might not have to register.

However, if you sell anything processed (such as sausages, burgers, bacon, curried lamb), you must register. And no matter what meat product you sell, you must also register if you do what is called 'selling at a distance': this is defined as anything other than selling face to face, and includes advertising your meat on the internet, whether it be on your own website or via social media, or selling your goods by phone. And you must register if your meat is transported or couriered to your customers, no matter how small the quantities involved.

In all cases, get in touch with the local EHO so they can determine whether you are exempt from registering or not. The good news is that registering is both quick and free.

A small meat enterprise that has all their meat cut and packed by a licensed butcher on the butcher's premises, with the keeper (or a courier) delivering the produce direct to the customer and never actually handling the raw meat, is seen as a low-risk enterprise. In this case you won't have to set up a meat processing room that complies with all the regulations because the butcher has taken on that role.

If you sell direct to the consumer you must put a statement of the weight in with your meat and what it is (pork, beef, goat, lamb), so make sure that your weighing scales are calibrated accurately, or get your butcher to label it. If you sell cured meat or charcuterie the label must state the presence of additives and preservatives, and if you sell processed

Vac-packed and labelled beef.

Vacuum-packed home-reared bacon.

meat such as sausages or burgers you must also include the percentage of meat in the product.

The best source for precise and current information regarding meeting food hygiene and labelling regulations, in addition to the EHO, is the Foods Standards Agency website. EHOs recommend that anyone starting a food business undertakes food hygiene training to ensure that they understand all the food hygiene risks.

SLAUGHTERING POULTRY AND WATERFOWL FOR HOME USE

The equipment and techniques used to dispatch birds can be harmful or lethal to humans. In no circumstances can the author or publisher accept any liability for the way in which information contained in this book is used, or for any loss, damage, death or injury caused, since this depends on circumstances wholly outside the writer's and publisher's control.

Freshly dressed duck weighing 2.07kg.

Don't overface yourself by planning to slaughter and dress too many birds in one morning. It takes me twenty minutes to dry pluck a chicken, an hour to do a duck, and two hours for a goose. To minimise the need for removing pin feathers, slaughter birds when they are in full feather and not moulting as they develop adult plumage. The earliest age to slaughter geese is sixteen to eighteen weeks, chickens at five to twelve weeks, and ducks at around ten weeks (for Muscovies estimate fourteen to sixteen weeks). That said, our preference for ducks is to wait until they have their adult feathering (at sixteen to twenty or more weeks).

We frequently take chickens and ducks up to twenty to twenty-four weeks to produce a bigger, slower growing bird without stuffing them with feed, and geese are raised to twenty-four to thirty-six weeks. Our older, home-grown birds are a lot less fatty than the bought, oven-ready birds of my pre-farming days. Slaughter in the morning after the birds have fasted overnight, so that you minimise the risk of bird faeces contaminating the carcase as you work.

Methods of Dispatch

Acquaint yourself with the *Practical Slaughter of Poultry* guidelines produced by the Humane Slaughter Association (HSA): this is available online, complete with videos and useful images. As the HSA says:

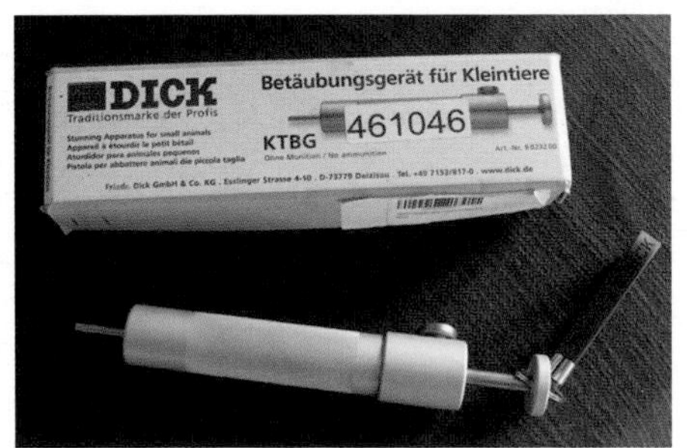

Stunner suitable for rabbits and poultry up to 16kg.

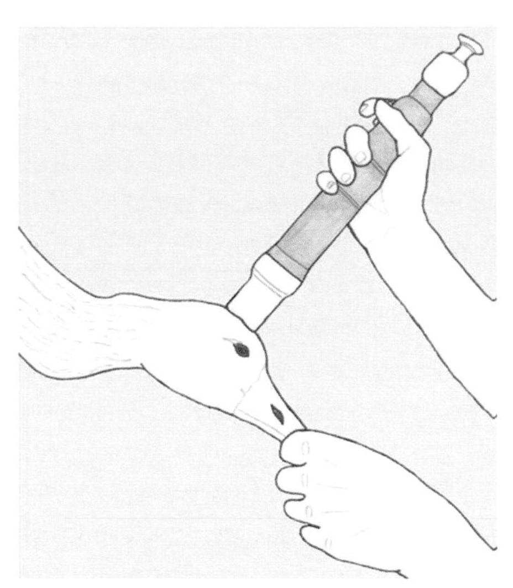

Correct positioning for concussion stunning of poultry (HSA).

'Slaughter may never be pleasant but it can, and must, be humane. It is your responsibility to ensure that you are fully prepared in order to protect the welfare of each individual bird.' You need to remain confident, patient, calm and quiet when slaughtering birds; this means a knowledgeable confidence, not a gung-ho let-me-at-them approach.

If you have any doubts about your ability, you must not attempt to slaughter any bird. Once stunned and killed, to ensure food safety, birds should be kept suspended for two minutes to allow the blood to drain from the carcase before plucking and evisceration begins.

Plucking – Dry and Wet

This is not a job to do in a draught or outside on a windy day, as fluff, down and feather will go everywhere, including up your nose. Nor is it a job for inside the house, as the mess will be prodigious. A shed or barn with a beam for hanging the birds head down, feet caught in a noose of baler twine, is the way to organise this task. If you do it sitting, be prepared for being covered from head to toe in feather and down. Pluck the birds while they are still warm, so only slaughter as many birds as you intend to process immediately: once the bird is cold, plucking is harder and you are much more likely to tear the skin.

Manual dry plucking gives the best quality carcase, and if you want to save the feathers, is also the method to use. The swifter alternative is wet plucking, which involves scalding the carcase by immersing it in water at 60–68°C for up to three minutes to loosen the feathers. An electric wash boiler or a large tea urn no longer used to make tea are ideal, or a very large saucepan of the kind found in commercial kitchens. You can also use a steam cleaner, wallpaper stripper, or an old steam iron to loosen the feather.

To avoid tearing the skin and spoiling the look of the carcase, pull feathers one or two at a time, not in handfuls, and pull in the direction of growth. Remove the wing feathers first, followed by the tail feathers, then move down the body, plucking well down the neck to the head. Plucking machines are efficient but expensive, though may be worth investing in if this is a regular activity.

In the case of waterfowl, once the feather has been removed it's time to start on the down, and any pin feathers. You can either do this by hand, or in a wax bath where the bird is dipped into melted food-grade poultry wax, then cooled in cold water, and waxed a second time. You then pull the wax off the bird, removing the final bits of fuzz. The wax can be warmed, filtered and used again. If not using wax, and for all birds, finish off with a kitchen blowtorch: light it, and pass the flame over the bird to singe off the last vestiges of fluff and any hairy bits.

Oven-ready turkey (with Berkshire streaky bacon).

Gutting

Birds can be drawn (gutted) as soon as they have been plucked, or hung in a cool place for up to a day. This hanging will produce a stronger, gamier flavour that may or may not be your preference. We choose

The Gutting Process

Remove the feet	Cut through the hock joint above the feet with a sharp knife or poultry shears.	*Removing the feet.*
Remove the head	Cut all the way through the neck as close to the head as possible with a sharp knife or poultry shears. For waterfowl you may choose to remove the end wing joint and discard as there's no meat worthy of note on this bit.	*Feet and head removed.*

Extract the neck	You'll want the neck for giblet gravy so cut the neck, skin and all, close to the body and then roll the skin off the neck. Remove the windpipe and oesophagus attached to the neck and discard. Put the neck and skin aside. If the bird is empty, you can ignore the hen/duck's crop (geese don't have one), but if it has food in – this will be obvious in the neck cavity – use your fingers to loosen the crop from the surrounding membranes attaching it to the body and cut it off as low down in the cavity as you can. Discard.	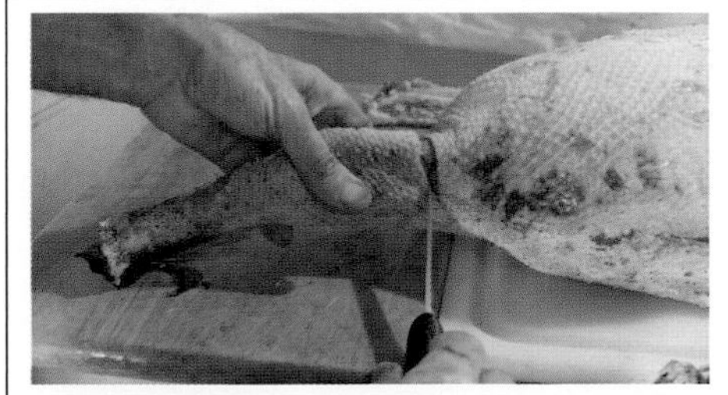 *Cutting off the neck.*
Remove the innards	Make an incision all around the vent, angling the blade away from the intestines so you don't cut into it. Hold this 'plug' and pull it gently away from the body with the intestines attached. Run your fingers inside and around the body cavity to break any connecting membranes and loosen any fat deposits (you may wish to keep the fat); the membranes are quite tough. Don't grab and pull at the intestines as they will break. Hook your fingers up and over the gizzard (a hard muscular ball) and then pull – the intestines will come out in an unbroken whole. Push this mass to one side away from the bird. Go back inside to extract the liver. Attached to the liver is the gall bladder, a green broad bean-sized organ – do not break this as it contains bile – and put carefully to one side. Go back in again to remove the heart and put aside. Check one final time for any odd bits of tubing such as the windpipe, and discard. Be aware that the windpipe is connected to the voice-box (syrinx), and pulling on it may make the carcase squawk (don't panic – the bird is definitely dead!).	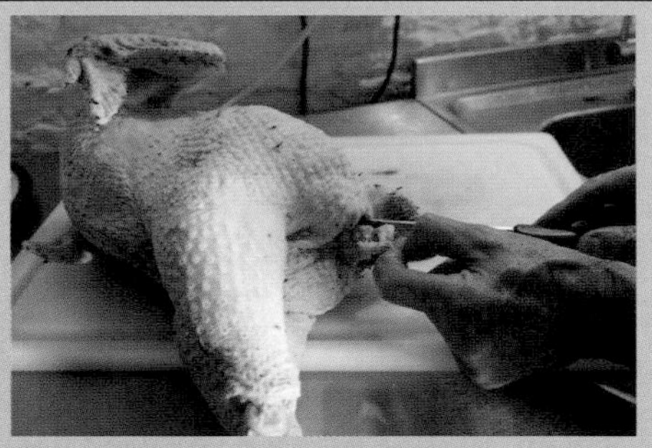 *Cutting around the vent.*  *Extracting the guts.*
Offal	Hold the liver carefully in one hand, gall bladder facing down, and with the other hand cut the gall bladder away from the liver, cutting into the liver, not the bladder, so there is no danger of spilling the bile. Discard the gall bladder. Take the gizzard and cut the intestines away and discard. Scrape any membrane and surplus fat off the gizzard and cut round the seam leaving a hinge. Open it up and you'll find grit and possibly grass: discard this. Peel away the thick yellow lining and discard that, too. The lungs can be removed or left inside the cavity; when the bird is cooked this is a treat for dogs or cats. The gizzard, liver, heart and neck are now ready to go in the stock pot or be bagged up for the freezer. The neck skin (for geese or ducks) is ready for stitching and stuffing or freezing.	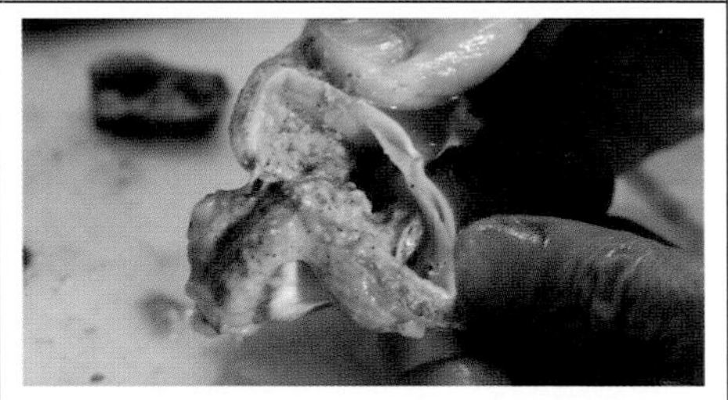 *Inside the gizzard.*  *Neck, heart, liver, gizzard.*

to gut our birds as soon as they are plucked. Have a rubbish sack close to hand for the disposal of all the inedible bits. It is best to do this on an easily disinfected surface, ideally stainless steel, next to a sink. Both plucking and gutting are fairly physical jobs.

You'll need the following items of kitchen kit:

- Large chopping board used specifically for raw meat and poultry
- Sharp butcher's knife
- Sharp kitchen scissors
- Cloth or kitchen paper for cleaning up as you go along
- Disposable gloves if you prefer not to go in with bare hands
- A tray for the bird
- A pot/bowl for the giblets
- Rubbish sack
- Apron.

You may prefer to remove the wing ends and add them to the giblet stock for gravy. We never bother removing the oil gland in the parson's nose, although some butchery guides suggest this.

Do *not* wash out the carcase (unless you've contaminated it with faecal matter), as all this does is risk the spread of bacteria all over your sink and worktop surfaces. Proper cooking, or freezing, defrosting and cooking, is all that is required. If you have spilled faecal matter on the bird, remove any gobs with kitchen paper. If you have made a worrying mess and really need to wash, do it from the neck end downwards, and be sure to cook the bird thoroughly; you will need to be particularly scrupulous in disinfecting the sink and surrounding area afterwards. Cool the bird properly in a cold room before putting it in the fridge or bagging it for the freezer.

If you're freezing the bird, put the giblets inside the cavity, and label with the weight, date and contents before putting it in the freezer. The Food Standards Agency recommends that whole birds should be frozen for no longer than a year, and pieces for no longer than nine months before defrosting, cooking and eating – but we have had birds in the freezer for years for personal consumption, and as long as the bag isn't ripped, which causes freezer burn (unsightly but not harmful), the birds come out as good as when they went in.

Access to Poultry Abattoirs and Slaughter on your Holding

It's increasingly difficult but not impossible to find poultry abattoirs willing to take a few birds to kill and dress for private customers, depending where you live. You may be able to find licensed slaughterers to come to your holding and slaughter birds for your own use. Keep up to date with any changes in the law by visiting the Food Standards Agency website (or the national advisory body in your location) and searching for 'Home slaughter of livestock'.

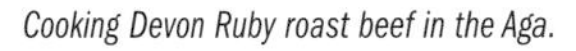

Cooking Devon Ruby roast beef in the Aga.

15 Grassland Management and Fencing

THE ENVIRONMENTALLY CONSCIOUS SMALLHOLDER

There is undoubtedly a danger that someone in possession of a small patch of land will try to wring every possible ounce of productivity out of it. It's understandable that smallholders, keen to try their hand at every potential activity, will not only burn themselves out, but tire out the land with overstocking. There is a desperate need for restraint, to allow nature its place, to feed, not starve the soil, while being appropriately productive. There is so much information out there that you can take advantage of, and determine if it holds lessons for your own land. Don't reach out for chemicals and heavy machinery as the first answer to a land 'issue'. From regenerative grazing to agroecology, broaden your horizons and see what resonates, and what can be adapted to the small scale.

FORAGE

For smallholders with ruminants, forage is the single most important crop that you will grow. Forage is plant material eaten by livestock – the silage, haylage or hay that is made from grass to feed livestock over the winter months. The difference between them is just a matter of time in their making.

When grass is cut it contains 75–85 per cent water. To make silage (60–70 per cent moisture) the grass is harvested quickly, just twenty-four hours after cutting: it is stored in clamps that once full, are covered with plastic sheets (frequently kept in place with a profusion of old tyres). For haylage the grass is left to dry for longer, and is turned to expose the damp clumps to the air. When there is still some but not too much moisture (around 30–40 per cent) left in the stems, it is baled and wrapped to stop it fermenting and going mouldy; cattle, sheep and horses love it. And then there is time-honoured hay, where the grass is left to dry even longer (16 per cent moisture) before baling; once baled it is stored under cover without any wrapping.

Hay keeps well, but wrapped haylage bales once opened will last only five to seven days before starting to rot, which is indicated by heat; once a bale warms any remaining haylage should not be fed but thrown on the muck heap. Unless you have a large enough flock or herd to eat large bale haylage, it is more economic to keep to hay. And it has the significant benefit of not creating so much plastic waste (eco-friendly options are emerging).

Haylage – winter forage.

Making Hay

As discussed in Chapter 5, if you want to make your own hay, you'll need a mower, a hay turner/rake, and a baler as well as a tractor. It's an expensive business with just a few acres, so calculate the cost of buying machinery used for a very short period each year versus the cost of buying in forage. You can increase stocking rates and buy in all your hay, or get in a contractor to do some or all of the haymaking for you, although with a small acreage you will not be their priority. If you do intend to make your own hay,

Making Hay

Mowing	It's summer (June or July if you're lucky), the weather forecast is sunny for at least a week, ideally the grass you intend to cut is just coming into flower but has not set seed. The machinery (or for the strong armed, your scythe) is in good order – you go and mow once the dew has dissipated off the field. First you mow up and down the field (clockwise) and finish off with the thicker grasses of the headland (the circumference of the field).	 *Preparing to mow.*  *Mown field.*
Tedding/turning	The cut grass will need spreading and turning to dry out. Do this once the dew is off the ground; you may do this once or several times a day depending on the weather. The hotter and dryer it is, the less often you need to turn the grass.	 *Tedding.*
Rowing up	In order to get the hay baled it needs to be collected into rows. Some tedders will do this, or you will need a separate rake.	 *Rowing up.*

Baling	Don't get ahead of yourself, baling while there is still dampness in the grass; if you store damp hay it will heat and the risk of spontaneous combustion is very real. You may choose small bales, large bales, round or oblong (but each baler only produces one type). Small bales can be carried by hand or in a wheelbarrow – perfect for the smallholder. Large bales require a tractor to move each one.	 *Baling large bales.*  *Baling small bales.*
Carting	Once baled, small bales should be stacked so that they can be thrown on to a trailer (or picked up by a flat eight or some other mechanised grab).	 *Bale carting.*
Storing	Hay needs to be stored under cover, and there are differences of opinion as to whether you sacrifice the bottom layer by stacking directly on the ground (turn the bottom layer bales on their sides if so, to suck up less moisture), or on pallets, which can create a cosy haven for rats. Your choice.	 *Storing hay.*

be prepared to be patient, and do not get frustrated because in the week you took off work to do it, it rains every day. For the determined, haymaking can be done by hand on a very small scale.

As a broad guide, expect five to ten large bales, or fifty to a hundred small hay bales per acre depending on the quality of the grass, the soil, weather and management. Hay can keep well for two to three years or even more.

When the weather is against you, wrapped haylage can be a saviour. The bales don't need to be kept under cover, and rather than a ruined hay crop and starving livestock in the winter, the damper mown grass can be wrapped. A wrapper is probably not in most smallholders' armouries, but a contractor can do this for you. Be aware that you will need to check on the wrapped bales regularly and swiftly patch any holes created by crows and the like, as keeping air out of the bale is essential to stop it rotting and going mouldy. Well managed haylage can be kept for a second year, but not much longer than that.

Hay (and straw) are available in a variety of bale sizes – those shown below are 'square' rather than round bales, the square (in reality oblong) being those available from hay merchants as they are more efficiently carried on transporters. Straw is the stalk of an arable crop (wheat, oats, barley are most common), which is baled up and used for bedding and feed.

Bale Weights and Sizes

Bale type	Approximate weight (kg)	Approximate size (metres)
Hesston	500–550	1.2×1.2×2.5
Quadrant	250–350	1.2×0.7×2.5
Mini hesston	240–290	0.8×0.9×2.5
¾ hesston	400–500	1.2×0.9×2.5
Small bale	18–20	0.35×0.75×0.9

Tree Hay

Tree hay is made by cutting branches from deciduous trees when they are in full leaf (in the summer), bundled up and hung in a dry place. This is a fairly labour-intensive process, but it produces nutritious browse for goats and cattle. Alder, ash, aspen, beech, birch, hazel, hornbeam, rowan, sycamore, willow and wych elm all make quality tree hay.

Muck spreader.

LOOKING AFTER THE SOIL

Manure: Black Gold

What do you do with all that muck stored in a steaming muck heap? That wonderful black gold is entirely an upside of having cows and goats inside for winter. The muck heap rots down over the year and is spread in late summer or early autumn, prioritising any fields used for haymaking, putting much needed nutrients back into the soil.

Avoiding Soil Compaction

Putting heavy machinery on soft or wet ground will compact the soil and kill off the microbes that aerate the soil and do good work to keep your pasture healthy, besides creating ruts and enabling topsoil to run off the pasture. Avoid this by not having heavy machinery on the land unless it's firm and dry. Better to miss a year of activity (for example hedge-cutting) if the weather and ground conditions are against you.

Harrowing, Rolling and Topping

Chain, spike or tined harrows pulled behind a tractor or UTV can be used in the spring to rake out the dead grass and moss – known as 'thatch' – to allow the new grass to grow. Harrowing will also spread out any mole hills, so that if you are intending to make hay from the pasture you are a lot less likely to be baling soil in with the hay, contaminating it with listeria. Any large heaps of dung will be advantageously broken up and spread across the ground.

Rolling will flatten mole hills and ruts, and also squash loose stones into the soil.

Harrowing.

Topping with a compact tractor.

Hedge laying.

Topping is predominantly a summer activity, cutting down the grass-seed heads with their thick stems that the livestock don't relish, and managing nettles, docks, rush and thistles where they are encroaching on the pasture. Topping is carried out once the livestock have been through the pasture so that you are selectively removing unpalatable growth. It encourages new grass growth and also helps to reduce lameness; the thick seed-head stems cause sore feet as they rub uncomfortably between the foot claws as livestock walk through the pasture.

HEDGES AND TREES

Hedges and trees provide vital shade and protection for livestock, habitat for birds, insects, bats and small mammals, and fruit and nuts for them and you. If you are lucky you will have plenty of trees and hedges on the land, and will be able to plant many more. Source trees and hedge plants from nurseries selling native varieties grown locally to minimise cost, ensure healthy stock, and which are appropriate for your location, and choose a mix of plants to maximise diversity and minimise risks from diseases targeting specific varieties. More cheaply, grow on tree seeds and self-seeded saplings collected from your ground.

If hedges aren't cut or managed they turn into lines of thin, poorly spaced and therefore vulnerable trees, which decreases the wildlife and livestock value of the hedge. Ideally hedges should contain full-sized trees in the mix. You don't want to cut hedges every year as most fruits develop on two-year-old wood, nor do you want to cut them to the same height each time as this causes them to go gappy and not regenerate. Cut hedges every three years at most – a maximum of a third of your hedges cut each year in rotation between September and February, the later the better. Cut them slightly wider and higher each year – about a hand's breadth (10cm) will make all the difference, and in an A shape, slightly narrower towards the top and wider at the base for the benefit of birds and small mammals. Cut in late winter if

Spring regrowth on a newly laid hedge.

ground conditions allow, to maintain as much fruit for as long as possible.

Every twenty years or so the hedge will need laying in the local style – these have significant regional variations – so even though you may be making the hedge a little bigger year on year, it will all be laid again before it encroaches on the pasture.

A large dead tree in the middle of a field provides a perfect habitat for invertebrates, owls and more, so don't cut it down. Tree felling is a legally controlled activity and a licence is usually needed if the tree, or a group of trees, has a volume of timber over 5 cubic metres.

GRAZING SYSTEMS

Your task is to balance the supply and demand of the grass on your holding, improving its soil structure and biodiversity and not over-grazing or poaching it. Not long ago I was asked this question: 'If we have five cows in a ten-acre field, would that work all year round, or could you put five cows in five acres for six months and then move them to the other five acres for the second half of the year?.' My response was that you shouldn't do either of those things: it's not advisable just to put cows (or any grazing animal) in a field and leave them there for six to twelve months – known as continuous grazing or set stocking – if you want healthy livestock and pasture. You need to move livestock regularly to allow the grass to grow, and not let them eat it too short. A few alternative methods are described below to stimulate further research.

Regenerative, Mob or Holistic Grazing

Many voices internationally are spreading the word about mob grazing, also described as holistic or regenerative grazing. This approach simulates the effects on the landscape created by wild herds such as buffalo and bison, grazing livestock intensively for short bursts in small areas and then moving them on every few days or even hours. It is a system where grazing is in one spot for a short duration, with a high density of livestock, followed by a long grass recovery period. The aim is to graze a third of the pasture, trample a third, and leave the last third standing. The tightly mobbed flocks/herds are contained in their allotted space using electric fencing.

The trampling creates organic matter to feed the soil structure, minimises compaction, as the long grass protects and builds the soil, feeds the animals effectively, and allows the grass to grow to its full potential. The top third of grasses has all the energy

Using cattle for conservation grazing.

and sugar that the livestock need, and moving the herd on quickly lets the grass regenerate and grow back fast.

To do this you need to be able to subdivide grazing areas into small paddocks, allowing the grass to recover fully and grow thick and robust with deeper roots before the herd returns to it (sixty days rest is common); this also increases the overall grass yield significantly. There are real benefits to wildlife and flora too, as areas left ungrazed for long periods support wildflowers, birds and insects. It can also reduce the use of wormers, as grazing tall grass reduces the risk of livestock ingesting intestinal worm larvae, and the extended rest periods help to break the parasite cycle. Mob grazing can increase grass yield by 28 to 78 per cent, compared with set stocking.

There is nothing to stop the smallholder adopting the regenerative approach if you can put in place portable fencing and water, whether this is for chickens in a portable ark moved every day on to fresh ground, or sheep, goats or cattle being moved across pastures with electric fencing. Read Joel Salatin and watch Greg Judy videos on the topic of small-scale rotational grazing: the latter suggests you can raise two steers on 5 acres using the principles of grazing the top third of the grass and moving them on.

Bale Grazing

Rather than putting supplementary forage in a feeder around which livestock will stand and compact and destroy soil structure, bale grazing is a method whereby hay is spread out on the ground to allow the animals to feed. This might be wasteful of forage, but degraded land can be revitalised this way as it adds organic matter and seed from the hay as well as dung.

Strip Grazing

Strip grazing relies heavily on electric fencing, where the fence is moved once or even twice a day, where livestock are moved across a field, eating in front of themselves and mucking behind as they go.

Rotational Grazing

Rotational grazing has some similarity with mob or holistic grazing, with livestock moved from one small pasture to another every few days. Again, after a paddock is grazed it is rested to rejuvenate the grass ready for the next grazing period, but paddocks are not rested as long as paddocks in the mob grazing system, and the sward height is lower.

Bale grazing.

Silvopasture

Silvopasture is the planting of trees in abundance to provide forage and shelter for livestock to the benefit of animals and trees; in particular it is practised in warm climates, and is of increasing interest in the UK.

IMPROVING THE SWARD

Controlling Pernicious Weeds

There will be some plants in your swards that you may prefer to be without, or at least not establish a complete takeover, such as soft rush, creeping thistle and nettles, depending where they are situated. Stopping them from setting seed by spending a bit of time on summer evenings wandering around cutting them down with a grass hook (or a strimmer in the case of rush) can have a surprisingly positive cumulative impact.

Ragwort is much enjoyed by the cinnabar moth, but if proliferating in grazing areas should be pulled up by the roots (wear gloves) and burned; do not leave them to wilt on the field as ragwort is toxic when ingested by livestock. Avoid disrupting the soil and exposing dormant weed seeds to minimise dock and nettle growth; where we have docks the dock beetle is active and stops them becoming widespread. We find that young nettle leaves and cut and wilted thistle and nettle will be eaten by the livestock; those summer evening wanderings, grass hook in hand, are to be greatly encouraged.

Soil Acidity

You or a local agronomist can test the pH levels of your soil and determine if it is acidic and would benefit from the addition of ground limestone or chalk. Spreading lime or chalk decreases the acidity of the soil and increases the number of worms, which in turn improves the soil structure. It also makes the grass more palatable to livestock, increasing the amount of magnesium and calcium in the sward.

Yellow rattle.

However, acidic soils do provide a suitable base for native wildflowers, and wildflowers actively thrive in poor soil where nutrients are low.

Creating Diversity

Grazing with livestock actively helps the biodiversity of pastures, through the action of feet, mouth, urine and dung. Following regenerative practices contributes significantly, but you can also plant in diversity by scarifying some areas of sward and casting native wildflower seed mixes; this might include yellow rattle, which is a semi-parasitic plant that feeds off the nutrients in grass roots, decreasing the amount of grass and leaving space for other, less vigorous plants to flourish.

FENCING

If the holding you have or are looking to purchase doesn't have good fencing, do include the cost of sorting it in your budgeting; you simply can't have livestock without it, and it's not cheap. Effective fencing enables you to segregate different groups of animals, protect growing crops, provides a degree of protection against some predators, and prevents your animals from straying on to neighbours' land or getting out on to a road with potentially disastrous consequences. You should always concentrate on making your boundary secure before moving on to internal divisions.

Even if you intend to pay a contractor to do fencing, you need to know what to ask for, and understand how it's put together so that you can do those ever-present repairs. There are many types of fencing available for different species and applications, and the two most useful are stock fencing and electric fencing. Stock fencing will contain sheep, cattle, pigs, horses, large poultry, waterfowl and most llamas and alpacas; goats, however, require some additional features. Electric fencing is a simple DIY job with the components readily available and no specialist tools required. Stock fencing is a little more complex, but with savings of 50 per cent to be made by doing it yourself (plus the satisfaction of admiring your own handiwork) it may be worth learning how to do it.

Stock Fencing

Stock fencing consists of posts and stakes (usually timber, but they can be galvanised steel) driven into the ground, and high tensile (HT) wire stock netting

Stock fencing.

Goat fencing.

stretched tightly between them. The netting is topped by two strands of barbed or plain wire to achieve a final height of about a metre. Standard stock fencing is 80cm tall with eight horizontal strands and vertical gaps of 15 or 30cm; for sheep and goats choose fencing with 30cm gaps (HT8/80/30) rather than with 15cm gaps (HT8/80/15), as the wider openings mean that heads and horns are much less likely to get stuck, and there is less risk of ear tags being ripped out.

As goats have perfected the art of escaping, if using standard stock fencing, use long fencing stakes (1.8m) to add three strands of wire or electrified tape to a total above-ground height of 1.3m. A line of rails attached to the stakes along the inside top line of the stock fencing is also advisable for goats, as they will stand on fencing. Lambs and even the biggest goat will wriggle under fencing, so make sure it's taut at ground level and doesn't have any enticing gaps at the base; this is not an easy ask on hillocky ground, and bits of rail may be needed to fill gaps at ground level. The essential bracing of fences using diagonal struts at field corners can offer a convenient jumping-out point for goats, so use a box strainer design, or ensure conventional strainers are behind the fencing, and not in front of it.

Space fencing stakes for sheep and cattle 3.5m apart, and much closer together for goats, ideally no more than 1.75m. Don't contemplate using half-round stakes for fencing as these aren't fit for poultry, never mind four-legged beasts; use 75–100mm diameter stakes.

For sheep and cattle bang in fencing staples on the bottom and top wires, as well as three of the interim horizontal wires, and for goats bang in staples on every strand on each stake. Put stock fencing on the inside not the outside of the stakes so that when livestock push against it, it won't be pulled free of the fencing staples. Ensure stock fence is 'twangy' tight, using a boundary fence clamp, hand pullers and wire strainer.

If you are only intending to have cattle in a particular field you can get away with three strands of barbed wire pulled taught from stake to stake. However, this renders the fence entirely useless for containing

Cutting a notch into the strainer post.

Banging the strut into place.

Box strainer.

Corner post.

Using the wire strainer.

sheep, goats or even young calves, so if in doubt about a parcel's range of uses, go for stock fencing.

Permanent pig fencing should use a thick gauge of stock fence wire (HT/8/80/15) plus a strand of barbed wire just above ground level to deter digging underneath the fence – pigs are far more likely to make their great escape through digging rather than leaping. Electric tape attached to insulators on the inside and standing proud of the stock fence at 15cm and 30cm is a great preserver of the integrity of pig fencing too, as long as you remember to keep the electric on.

If using timber, buy fencing stakes and posts that are guaranteed for fifteen years; the additional cost will be insignificant in comparison to the effort and cost of replacing every post within a decade.

Driving in large posts is best done with a tractor-mounted post driver; a hand post bumper is fine for small sections of fencing but is a daunting prospect

Using the boundary clamp and winch to tighten the stock fencing.

KEY FENCING TIPS

- Gates require a sturdy hanging post and closing post.
- Fences should not be strained to the gate hanger or closer.
- Stakes/posts go on the far side of the fence except at corners and on concave runs, where they are on the field side.
- Straining posts must be strutted in both directions of pull. Struts should be at an angle <45 degrees to the ground, notched into the post at the top and buried in the ground. Put a stone under the foot of the strut, or have it push against the first stake in the fence.
- A stock fence wants to go in a straight line horizontally: if the ground is very uneven you may need to do several short runs in order to follow the contours.

The process:

- Mark out fence lines and lay out posts.
- Drive in posts and turners.
- Hang gates.
- Fit struts to the end strainers. Fit any intermediate struts/wire ties.
- Attach the wire to the first strainer; wrap round the post, staple and pigtail.
- Roll the wire out, downhill if possible. Kick the wire out with your feet, and release any snags with pliers. Keep your fingers out of the roll.
- Attach the boundary fence clamp to the wire, with the nuts on the outside; pull the points inside. Attach the clamp to the hand pullers connected to the end post by chain or strop. Tighten the wire, easing on intermediate stakes as required.
- Ideally leave the wire to relax for a while.
- Tighten the wire fully. Pull the last piece of wire to the post using pliers/strainer, wrap it around and staple. Angle the staple almost horizontally to fully pinch the wire. Pigtail the ends.
- Staple the wire to the turner posts, then to the stakes. Staples should be just off vertical, driven in to just touch the wire, which should be free to move.
- Attach the first strand of barbed wire, roll out and strain. Staple at the end, then the intermediates.
- Repeat for the second strand.

Safety first:

- Be careful of heavy items – a 100m roll of stock fence weighs about 65kg. Lift it correctly, or roll the wire/posts to avoid lifting.
- There are chemicals in tanalised timber so wear gloves.
- Rolls of netting are tightly coiled and unwind rapidly when opened; do not get your fingers caught in the roll.
- When cut, wire whips back; even fencing that has been up for years has memory of being coiled and springs back. Be particularly careful when working in pairs.
- Wear thick leather gloves for barbed wire no matter how horny handed you are.

for all but the fittest when dealing with long runs. You should be able to find a local contractor prepared to do this; you can save money by having all the posts and stakes set up exactly as you want them before they arrive, and you can attach the stock netting yourself if preferred.

Steel Fencing Systems

An alternative to timber stakes are steel fencing systems that don't need a tractor-mounted post banger to erect. Stock wire clips quickly on to galvanised steel posts and has a long life.

Poultry Fencing

To fox-proof permanent poultry fencing it should stand 2m high, with an overhang projecting outwards from the top and an outward-facing apron of mesh dug at least 30cm below ground level. Sagging fencing becomes climbable by foxes, so keep fences taut. Consider chain link for enclosing very large spaces, while heavy-gauge weldmesh is useful in smaller areas, and can be put along the base of a hi-tensile or chain-link poultry fence to keep chicks, ducklings and goslings in and mink, weasels and rats out.

Avoid chicken wire as the main fencing material as it will eventually be breached by all chewing creatures from rats to foxes, and can't be pulled taut. Heavy-gauge 25mm chicken wire (aka hex netting) can be used as a cheaper alternative to weldmesh along the base of your main fencing to keep small creatures in (and out).

Clipex steel fencing.

Electric fencing kit.

Electric Fencing

Electric fencing can be used as the sole means of containing animals, as a backup or reinforcement to a solid fence and as a deterrent to predators. One or more strands can be run around the outside of a poultry enclosure to deter foxes. A single strand along the top of a fence will prevent goats and cattle leaning or rubbing on the fence.

In addition to electric fencing (preferably netting or tape as wire has the potential to be dangerous as it can wrap round livestock and cut into them) you will need an energiser, an earth rod and a power source: a leisure battery, solar panels or mains. Solar is a useful back-up for keeping the battery charged, but may not be reliable enough on its own in extended cloudy weather. Remember to check that the fence is on and working each day; pocket-sized electric fence testers are readily available, and hang a sign on the fence that explains to any visitors that it's electrified. Electric netting should not be used for horned sheep and goats or for geese as they can get entangled.

Cattle need a height of only 80cm and one or two strands of tape, and sheep require three strands from 70cm down. Pigs require a height of only 40cm (lower for piglets) and one or two strands of tape. For poultry, use the tallest poultry netting, which is 1.2m high, and strim the grass by the base of the fence regularly to stop it shorting out. Birds that are flighty should be wing clipped to stop them flying over the fence.

16 Making It Pay

A smallholding doesn't have to be a business. Many people keen to grow their own produce and rear their own livestock are nervous that they might have to incorporate a business ethic into a way of life that they see as gentler, healthier and more self-contained, without the demands of doing accounts, creating a platform for sales and dealing with customers. If smallholding is intended to be your escape from business and employment, relax into that and don't see yourself as in any way beholden to the concept of making it pay.

For some it will be important that enough income is generated so that the holding breaks even and doesn't become a drain on earnings, savings or pension, while for others the concept of making it pay may be critical to becoming smallholders in the first place. If the latter scenario is your plan, it's crucial to appreciate that the skills needed by any small business will also be needed here.

The production of goods or services for sale is just a small part of the whole. You will need to be an efficient project planner, book-keeper, website designer/creator, marketing director and copywriter, and be happy dealing with customer relations, supplier liaison, payment chasing and a whole lot more, including meeting all the legal requirements for selling. You can pay someone to do some parts of this for you, but then you'll need to earn even more to pay them for their skill.

On the positive side, there is a big and increasing demand out there for great produce of known provenance. It only takes a horse-meat scandal or public concern over industrial lab-grown 'meats' for the demand for our own meat boxes to go through the roof. Our meat-box business is tiny – we send out around 250 meat boxes a year – yet we have hundreds of customers and new people signing up to go on waiting lists every week. We are not increasing the scale of what we do, and our supply cannot possibly match the demand, but it is an indication that there is huge potential out there for plenty of new small producers – and that's just on the fresh meat front.

DEFRA DEFINITION OF A COMMERCIAL HOLDING

Commercial holdings are defined as those with significant levels of farming activity, that is, holdings with more than 5ha of agricultural land, 1ha of orchards, 0.5ha of vegetables, or 0.1ha of protected crops; or more than ten cows, fifty pigs, twenty sheep, twenty goats or 1,000 poultry.

It's not for me to tell you what lucrative business opportunities you could create – sometimes it's about offering a portfolio of goods and services, while for others it's specialising, focusing on one passion project.

I don't believe that constant growth is a good thing: we live in a finite world with finite resources, and stressing the planet and ourselves in a consumerist frenzy is both dangerous and unsatisfying. Doing what you do better, however, is definitely a good thing, so grow your knowledge, your interests and your happiness if you can. Certainly you can increase your yield by putting in costly inputs (artificial fertiliser, hard feed, mechanisation and over-stocking), but all you do is over-stretch your wallet and ruin your land.

When acreage is small and you want to try everything, you can overreach what your body, mind and soil can deliver. So step back, rethink, and add value to your end product rather than attempting unsustainable quantities. If you are losing a fiver on every sheep you own, trebling the size of your flock will just make things three times worse. Instead, make each sheep work for you: add value by using

Some Income Generation Opportunities for Smallholders	
Christmas turkeys and geese	Market garden
Point-of-lay chickens	Micro greens
Eggs – eating and hatching	Cut flowers
Sheepskins	Edible flowers
Fleece for spinning and weaving	Pick your own fruit
Knitting wool	Cider, vinegar and apple juice
Hay	Dog walking field
Honey	Kennels/dog day care
Livestock – from quail to cattle	Micro dairy (butter, milk, ice cream etc)
Alpaca trekking	Care farm
Horse grazing and livery	Forest school
Campsites/glamping/shepherds' huts	Wedding venue
Holiday lets	Mushrooms
Meat box scheme	Jams, relishes, pickles and preserves
Charcuterie and smoked foods	Artisan sausages
Willow for basket making	Charcoal
Educational visits and walks	Christmas trees
Pig share scheme	Halloween pumpkins
Fencing and land management	Trailer hire and livestock transport
Shearing	Foot care for sheep and goats
Firewood	Poultry huts

the wool, the horns, or, as in our case, selling glorious sheepskins that are a direct by-product of the meat.

BUSINESS OPPORTUNITIES

Selling Meat

There is huge potential for small businesses wanting to avoid the middleman (that is, not selling your animals at market, direct to the abattoir or to the local butcher) and instead selling directly to the end consumer. The more nervous that consumers are about meat, the more those who are determined to continue eating it are seeking it from small-scale producers. Provenance is key, plus the customer's ability to ask the smallholder exactly how it's reared.

Offering everything from geese for Christmas, pasture-fed native-breed beef and lamb, and products found less commonly at the butcher such as goat, hogget and mutton, has evergreen popularity. You could do worse than learning from some of the great charcuterie makers out there, adding significant value to the base product – perfect for a smallholding business. You might get £10 per kilo for sausages, but £25–£55 per kilo for salami and chorizo.

Going Beyond the Artisan Sausage

Who doesn't love a good sausage? Comforting, loved by children and adults alike, I am not decrying the lure of the superior sausage, but simply using it as an example of a possibly saturated market. One can build on tried and tested ideas, but being ahead of the game and creating new products or opportunities can be particularly lucrative for the entrepreneurial smallholder prepared to take a calculated risk. Not all of us have the instinct and push to create a new market, but some can, and I've seen some very successful examples over the years.

Campsites and Holiday Cottages

In an era where unnecessary air travel is becoming both environmentally unsustainable and philosophically uncomfortable, holidaying in the countryside of one's own land is more popular than ever. Whether it's holiday cottages, luxurious yurts, log cabins and shepherds' huts, or space for a tent and a few caravans, smallholders are offering holiday options to those in need of a break, often with the added fun of interacting (safely) with the livestock.

You usually need a camping or caravan site licence (from Natural England) and planning permission from your local authority to use an area as a campsite, but you can apply for an exemption. One way of achieving exemption is by becoming a Camping and Caravanning Club Certificated Site, permitting up to five caravans/motorhomes, plus a maximum of ten tents, to stay for a period of twenty-eight consecutive days at any one time.

Holiday lets are another smallholder option, with conversions of barns and outbuildings. These must be furnished and available for letting for at least 210 days a year, with any single let no longer than thirty-one continuous days. There are a number of health and safety requirements and you will need to check with your local authority whether planning permission is required.

Farm Shops

Smallholdings are often situated in the most out-of-the-way places, but perhaps you are located on a busy thoroughfare with plenty of passing trade. If so, do you have enough produce to warrant having a farm shop, or can you create a network of like-minded smallholders to supply a wider range of goods? Whether you open for a couple of days a week or are a key retail fixture, a farm shop gives people somewhere to come and see your goods and potentially make more purchases than envisaged.

Planning permission for a farm shop can be secured either by making a planning application, or through permitted development rights that allow change of use of agricultural buildings to a flexible commercial use when certain conditions are met; in either case you need to talk to the planning department of your local authority.

Farm Visits

Educational and other farm visits are very popular, and the easiest way to start is to participate in national schemes such as those run by LEAF (Linking

Adding value to the produce stall.

Eggs for sale.

Environment And Farming), who provide all the practical information needed, from risk analyses, handwashing stations, boot dips, parking, insurance, and a whole lot more.

MARKETING

There are many hefty manuals and whole degree courses on marketing, so I'll keep this limited to some very simple truths.

- If the very notion of talking about your produce to prospective customers makes you panic, don't even contemplate selling – keep to producing for your own table.
- No one will buy your produce if they don't know it exists. You have to get the message out there.
- Reaching your potential customers may cost you very little money if you are canny. A good use of social media, a genuinely attractive, informative and easy-to-find website, and an ability to share your story, can be far more effective than paid advertising.
- A stall at your local farmers' market can be a great help in getting your products known and sold; just make sure the produce looks fantastic and the edibles good enough to eat.
- As regards the product itself, foodstuffs need to look and smell great, be presented well, and make the customers delighted when they eat it. Non-edibles need to look desirable, so no raggedy lambskins or knitted goods that only a fond relative could love.
- Whether selling on-line or face to face, make sure your prices are obvious to every prospective customer. If they can't see your prices or need to phone or email to find them out, you've lost them.
- Make sure pricing gives you a return. Understand your costs, know what competitors are charging, and don't subsidise other people's suppers or lifestyles. A business that makes a loss on every item it sells is unsustainable.
- You have to love your own produce. People who enthuse about what they produce are fun to talk to, informative and convincing. People enjoy buying direct from the smallholder.
- Think about your packaging and keep it cheap, minimal, simple, and as environmentally friendly as possible. Fancy packaging makes people think they are paying over the odds and contributing to yet more landfill.
- For foodstuffs, recipes are always welcomed by customers; add some inspiration to their meals.

Finding your Customers

The usual (and effective) route for smallholders is selling to family, friends and neighbours. If what you produce is of quality, word soon gets round and friends of friends start to buy too. A good website, and an active social media presence announcing when things are available, means you may never have to stand behind a market stall, unless that's something you'd love to do. An online presence is critical for reaching customers beyond your immediate network; make it informative, friendly and to the point, with product availability and prices made clear. Keep it as straightforward as possible for people to buy your goods.

Plan how you intend to distribute to customers, whether it's at the gate, the farmers' market or by post or courier; for the latter do your homework, as couriers are not all equal in reliability. Prospective customers can be put off buying if the courier you use has a poor reputation.

Your produce has to look and be good, be presented well, and be exactly as described if you want repeat sales. There are many ways to sell your produce, and some approaches will suit you, and others won't: choose what makes you happy and gives you a buzz. If spending every Saturday morning at a busy local food or farmers' market doubles as your social life, you're on to a winner; if attending food festivals and food halls at agricultural shows or craft fairs appeals, then go for that, or approach retailers who pride themselves on stocking local produce.

Some Legal Requirements

There are legal requirements for selling milk and dairy products, cider, jams and chutneys, and eggs. The Food Standards Agency (FSA) is responsible for ensuring food safety and food hygiene across the UK. They produce numerous online guides on meeting necessary regulations, and local authority Environmental Health and Trading Standards departments also publish guidance leaflets. There is a range of regulations you need to abide by when selling food and drink, cosmetics and skincare, and they depend on whether you sell direct or to another retail outlet. You must register with Environmental Health as a food business twenty-eight days before selling produce.

For cosmetics made perhaps using honey, milk, herbs and other home-produced ingredients you need a valid safety assessment and certification. Trading Standards and Environmental Health officers can give you advice about trades descriptions, labelling, weights and measures, batch numbers and traceability, hygiene requirements and more on the Chartered Trading Standards Institute (CTSI) website.

Selling Cider

Cider is legally defined as 'cider or perry of a strength exceeding 1.2 per cent alcohol by volume (ABV) but less than 8.5 per cent ABV obtained from the fermentation of apple or pear juice without the addition at any time of any alcoholic liquor or of any liquor or substance which communicates colour or flavour other than such as the Commissioners may allow as appearing to them to be necessary to make cider or perry.'

A cider maker can make up to 7,000 litres of cider within twelve consecutive months before having to register for duty. This is not an automatic exemption, and if you intend to make any real quantity you need to apply for exemption through HMRC.

Selling Jams and Chutneys

If you are selling jams and chutneys you must register with Environmental Health as a food business. Labelling and composition guidance is available from Trading Standards.

Selling Eating Eggs

The normal regulations for selling your own eggs do not apply in full to hen eggs sold directly to the consumer for their own use if you are selling on your holding, or selling door to door, or at a local market. However, you have to avoid using legally defined terminology, including the quality or weight grading terms applied, for example, in shops: you cannot use terms such as small, medium, large, very large, free range or barn, as these have legal definitions.

If you have fewer than fifty birds and sell at a public market you do not have to mark your eggs with a producer code, but must display your name, address, the 'best before' date, and advice on how to keep eggs chilled after purchase (many people wouldn't dream of keeping eggs in the fridge, but there you are).

Farmers' Markets

Goods sold at farmers' markets must be produced by the person selling the goods (or their family or employee actively involved in the production), and local, or at most regional production is required. Individual markets have their own rules to ensure the provenance and appropriate nature of what is allowed on sale, and some market associations will inspect your holding. Selling food at farmers' markets still requires you to register as a food business, and to meet all the relevant food hygiene and safety requirements, labelling, weights and measures, and to have appropriate insurances in place.

The Farmer Retail Association (FRA) supports farmers and smallholders who want to sell their products direct to the consumer.

THE COST OF PRODUCTION

Pricing

A critical element of selling is understanding the cost of production. It is simply unsustainable to price things too cheaply. Keeping a careful record of your costs is essential for knowing the minimum price you can charge. You might decide that including your time is a step too far (which wouldn't be the case if you were employing someone else), and that capital items such as tools, equipment and building adaptations will last you many years, and only a percentage should be taken into account. That wouldn't be sustainable for a business, but may be acceptable if smallholding is a hobby.

However, you can't hide away from including the costs for consumables such as feed, medication, vet bills, slaughter, butchery and packaging (if we're talking about meat sales). If it costs you £3.50 per kilo just to have your lamb slaughtered and butchered, not considering any of the costs you've incurred in breeding, rearing, feeding and caring for that lamb (estimate at £30), selling 12kg of butchered lamb to a local bistro for £5 per kilo is illogical, as you'd need to charge £6 per kilo just to break even.

Selling direct to the consumer means you take control over pricing as long as it's competitively positioned. You'll need to do regular research to benchmark prices against others selling the same or similar products. When benchmarking against established producers and retailers it's not just price similarities that need to be considered: the quality of the product and the buying experience has to be equal. Customers consider more than their budget when shopping: eyes, ears, taste buds, emotions, politics and principles all come into play in making a purchase.

Critically, make the prices very easy to find: very few people are going to email or phone you to ask how much your products cost, and if it's on a market stall the law requires clearly displayed prices.

Costings

The moment a cost is written down it's already out of date, but that doesn't mean you shouldn't be keeping a close eye on income and expenditure. I keep a spreadsheet for each species of animal on the farm, and know exactly how much we are spending and the correlating income for every year; it's the only way I know for sure if keeping our chickens is a drain on resources, and whether the sheep are genuinely making a financial contribution. It alerts you to the exact increases, for example, in bought-in feed, which fluctuates almost daily depending on global wheat prices, and it is easy to compare years and analyse how the changes you're making to your activities impacts financially.

For years I sold hatching eggs, which not only paid for all poultry feed, the materials for making huts and runs, and adding to the collection of incubators, but made a profit and provided us with 'free' eggs and meat. All the same, it was massively time-consuming for a relatively modest return, and with the arrival of a small herd of cows the slack was more than taken up by the introduction of beef-box sales. However, chickens are forever popular, it isn't outrageously expensive to get started, and new keepers seem to come on stream every year; if you have quality stock and are adept at breeding, selling numbers of young birds and hatching eggs can be an attractive proposition.

There are publications produced annually and bi-annually (*see* Further Information in the appendices) that give up-to-date costings models for every imaginable area of agriculture from rearing wild boar to strawberry growing and holiday cottages. If generating an income from your smallholding is a necessity, do your homework and budgeting, and be thorough about costs and realistic about income.

Grants

Don't assume that the public purse will subsidise your smallholding activity. The availability of grants has diminished over the years. Some grants have a minimum 5ha (12.35 acre) requirement for eligibility, and require active farming of the land. Your best bet is to make contact with organisations that support activity that improves local nature recovery, environmental sustainability and water quality, such as the local Wildlife Trusts, The Woodland Trust, The Rivers Trust, People's Trust For Endangered Species, national parks, the Forestry Commission, Natural England and the Rural Payments Agency.

Public money comes with many responsibilities, which not only includes doing the work you've signed up for, but keeping written and photographic records, and being inspected; if you can't cope with bureaucracy, this is not something to enter into lightly, as the administrative burden of both applying and complying can be significant. On the other hand, we have been able to carry out environmental improvements that we could not have managed without grant support.

VALUE ADDED TAX (VAT)

VAT is the tax levied on the sale of goods or services by businesses, collected on behalf of HM Revenue & Customs (HMRC). The business then pays VAT to (or reclaims VAT from) HMRC by calculating the amount of VAT charged to customers, minus any VAT they have paid on their own purchases. All goods and services are either VAT rated or VAT exempt. The current standard rate of 20 per cent applies to most items, with a reduced rate of 5 per cent applied to a few items, and a zero rate applied to many items that are relevant to the smallholder.

You must register for VAT with HMRC if your turnover is more than the threshold (£85,000 in 2022/23). That might seem quite an ask for most smallholdings, but you would be well advised to register for VAT voluntarily (if you qualify) as there are significant financial benefits to a smallholder business. Most of your business expenditure will be standard rated for VAT so you will be paying VAT for the goods you buy. However, smallholding and farming are unusual businesses in that most sales are likely to be zero rated, which means you won't be adding VAT on to many of your sales invoices. If you are VAT registered, voluntarily or otherwise, you claim back all the VAT you paid that is over and above any VAT you charged.

This can amount to a great deal of money, particularly when you are acquiring expensive items such as machinery, equipment, vehicles and tools in the early years to get your smallholding up and running. That small tractor you bought, the pig ark, a shed for the goats, the milking machine, all those hurdles, troughs, hayracks and fencing tools: the going VAT rate percentage of the cost of those purchases will come back into your pocket. And if you're erecting or restoring a barn or putting up a lot of fencing, just think of the savings. If you are not VAT registered then you won't be able to reclaim any VAT.

The qualification is that you do have to generate an income to be VAT registered – it can't be all about clawing back outgoings without there being incomings. If your smallholding is a hobby it is unlikely to be a business for VAT purposes (if you're unsure, check the HMRC guidance on www.gov.uk). Generally, if you are providing goods or services in return for a charge, there's a business activity for VAT purposes. You will need to keep your smallholding accounts separate from your domestic finances, but in the age of computer spreadsheets that shouldn't be much of a burden; it is also hugely helpful in checking if you are losing money, only covering your costs, or are making money and have a viable concern on your hands.

If you move to your smallholding thinking that no income will be forthcoming, but a couple of years on realise that you are making sales, you can register for VAT and then backdate claims for VAT paid before registration: the time limit is four years for goods you still have, or that were used to make other goods you still have, and you should reclaim them on your first VAT return.

Zero-Rated Smallholder Items

'Zero-rated' means that goods are still VAT taxable, but the rate of VAT you charge customers is 0 per cent. The sale, hire or loan of live animals used in the UK as food for human consumption is zero-rated, including meat animals, dairy animals, poultry (not ornamental breeds), including those for egg production, honey bees, and rabbits (except ornamental breeds). Alpacas and llamas are standard rated so you do pay VAT for camelids, and charge it too, if you are VAT registered.

Aside from livestock, animal feedstuffs and nutritional supplements are also zero-rated, whether they are straights (such as corn, wheat, barley, oats), compound feeds (such as sow rolls, ewe nuts, layers pellets), or supplements such as mineral licks and rock salt. Hay and straw used for feed (for agricultural businesses only) are also zero-rated. If you supply grazing for animals the supply is zero-rated as it is perceived as animal feed.

Food and drink for human consumption is usually zero-rated, so any meat you sell will not incur VAT.

All crops specifically grown to produce food for human consumption or animal feed are zero-rated, including seeds, seedlings, herbs, crowns, spores, tubers and bulbs of edible vegetables and fruit.

Bartering and VAT

Please don't lose the will to exchange goods with fellow smallholders, but you do need to be aware that if you and your barter partner are registered for VAT you must both account for VAT on the amounts you would each have paid for the goods or services if there had been a purchase instead of a barter.

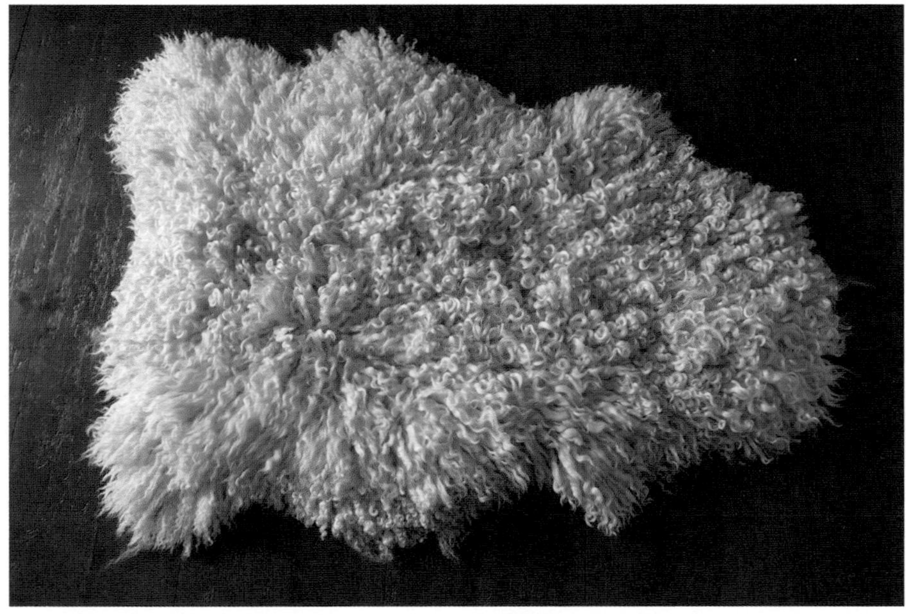

Whiteface Dartmoor sheepskin.

17 Making the Most of Your Produce and Resources

There are fat tomes written about each individual topic covered in this chapter, and a book on smallholding, no matter how substantial, can't do justice to all the potential ways to preserve and make the best of all your produce and natural resources. All the same, it seems only fair to share a few ideas that might intrigue, and a few that might point you in unexpected directions. This chapter peeps into the possibilities, from chutneys to firewood, vinegar to solar energy, butter to spring water.

Relishes and chutneys are an abundant feature in our scullery, and jars of their goodness get spooned into daily; bacon and chorizo make a regular appearance in the dishes we make; vinegar is as vital as salt in the kitchen; and the freezers are stuffed with edible produce.

For the tentative reader who wants precise detail as to a recipe so that it comes out perfect every time, I have sympathy, but my reality tends towards the 'I think this will work, let's have a go' approach, particularly when it comes to all things edible. If you swap one berry for another, throw in more beans and fewer onions, use gin rather than vodka, or adore or loathe garlic, go with your own flow: what's the worst that can happen?

LIQUID PLEASURES

We've made plenty of foraged elderberry and blackberry wine in our time, and make cider every year. Sloe gin, the ubiquitous smallholder tipple, reminds me forcefully of a particularly acrid brand of cough medicine, so the impressive number of sloes growing on the farm get left for the birds and insects. Elderflower champagne is a heavenly thing to make and drink: the room in which you make it smells like a joyous fairytale complete with happy ending. It's all about experimenting with the things that grow around you and deciding what you like best.

Elderflowers.

Apple Juice and Cider

Vats of cider are particularly important to us, as cider is wonderful used as a cooking liquor for pork dishes as well as serving in glasses to appreciative guests; but most of all it is excellent for turning into cider vinegar, which in one form or another we use almost daily.

But before you can make cider or apple cider vinegar, you need to make apple juice.

Making Apple Juice

To make apple juice you need the following:

- Apples. You can use dessert and/or culinary apples for juicing, but avoid cider apples unless you intend to make cider from them.
- A mill/crusher/scratter, for crushing/chopping the apples to produce a pulp from which the juice can be extracted.
- A press, for compressing the apple pulp to extract juice; both mill and press can be hired, and rural community groups have apple pressing days.

Bottled apple juice.

Milling the apples.

- Plastic (food grade) containers to collect the juice.
- Bottles or containers for storage.
- A large funnel with a nylon sieve to strain the juice before bottling.
- A small funnel for pouring into a jug for filling the bottles.
- Sterilising solution for cleaning equipment.

Method:

- Pick apples or use windfalls in good condition.
- Wash the apples and remove any debris. For juice, discard badly bruised or slug-infested fruit. For cider, bruises won't hurt, but I prefer slug-free cider.
- Mill the apples.
- Put the resulting pulp through the press.
- Collect the juice and pass through a sieve or muslin to remove any bits.
- Fill washed/sterilised containers with juice.

4kg of pressed apples, depending on the variety, makes approximately 2ltr of juice.

Preserving Apple Juice

There are two methods for preserving apple juice on a domestic scale: freezing and pasteurisation.

- Freezing is the simplest; use plastic bottles (sterilise plastic milk bottles, or you can buy new from smallholder suppliers). Leave a little space at the top to allow for expansion of the liquid when it freezes. The juice will keep its colour and flavour for several years.
- Pasteurisation is only successful as a means of preserving the juice if the whole container

Rinsing the apples.

Milled apple pulp.

Pressing the pulp.

Pomace.

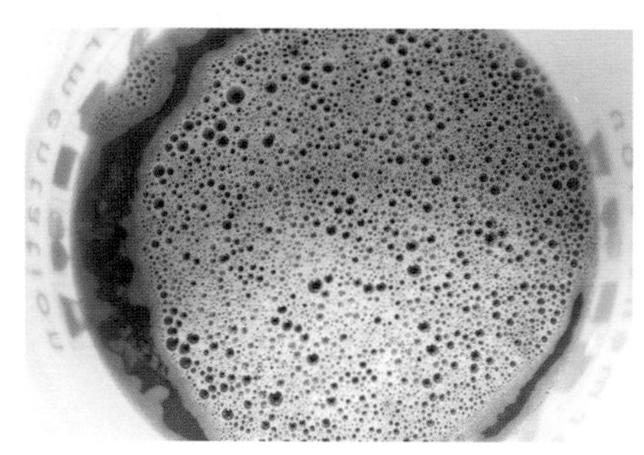

Apple juice.

is immersed. Pasteurising the liquid separately then pouring it into clean bottles is pointless, as the liquid picks up yeasts and bacteria that will create fermentation. In-bottle pasteurisers are available from home brewing and cider supply outlets. Once pasteurised in the bottle your juice will keep for one to two years.

Cider

There are four classes of cider apple based on their acid and tannin levels: sharp, bittersharp, bittersweet and sweet. Dessert apples lack tannin, but its presence in cider gives richness and body. You can use almost any apples to make cider, but do try to include some cider apples, use a mix of varieties, and avoid excessive numbers of very sweet dessert or very acid culinary apples.

Cider Apples

Classes of cider apple	Acid %	Tannin %
Sharp (high acid, low tannin)	>0.45	<0.2
Bittersharp (high acid, high tannin)	>0.45	>0.2
Bittersweet (low acid, high tannin)	<0.45	>0.2
Sweet (low acid, low tannin)	<0.45	<0.2

Making Cider

To make cider you need the same items as for making apple juice, plus:

- Barrels or buckets to hold the juice while it ferments.
- Airlocks to prevent oxygen or flies spoiling the brew.
- Bottles or plastic containers to hold the finished cider.
- Lids/stoppers/crown corks and crown capping tool for fixing the crown corks on to the bottles.
- Sterilising solution.
- A siphon tube.

You may also want to use:

- Campden tablets.
- Commercial wine yeast/nutrient.
- A hydrometer to measure the potential alcohol level.

Method:

1. Make apple juice as described. If you have unknown varieties, sample them as you go along to get some idea of how sweet/bitter/sharp they are.
2. Put the juice into plastic barrels or containers for large quantities, or demijohns for smaller amounts.
3. You can allow naturally occurring yeasts to develop and start the fermentation, or kill these off and use a commercial wine yeast. Using the natural yeasts is traditional, cheaper and, from our experimentation, produces a better result. If you want to inhibit the natural yeasts add crushed Campden tablets at the rate of one per gallon, leave for a couple of days, and then add a commercial wine yeast.
4. To measure the specific gravity at this stage you'll need a hydrometer to measure how much sugar is present so you can calculate how much alcohol the finished brew will contain.
5. Put the container in a reasonably warm place. The initial fermentation may be quite vigorous and cause froth to come up out of the container, so don't put the airlock in for a few days – use a bung of cotton wool, mop up any leakage, and replace with a clean bung as necessary. Once the fermentation has quietened down, put in an airlock. The airlock must contain water to work; crush a Campden tablet into the water to help keep it hygienic. Wild yeasts may take several days to get started; if nothing is happening you can always add commercial yeast.
6. Rack off after six weeks into a clean container, and then leave to ferment for several weeks/months. The length of time will vary depending on the vigour of the fermentation, the temperature, the amount of sugar in the liquid at the beginning, and the amount of sugar you want left in the finished product. Wait until the fermentation has slowed or stopped, then bottle it.
7. Once the cider is fermented you need to store it. Air is the enemy, so if you start to empty a large container the rest is likely to spoil, as once fermentation has ceased there is no carbon dioxide being produced to act as a preservative. We store our cider in bulk in the fermentation barrel, decanting it into bottles one barrel at a time.
8. If you want slightly sparkling cider you can add a teaspoonful of sugar to a beer or pressure-tolerant bottle and then fill with cider. This will cause a secondary fermentation and the resulting carbon dioxide will be absorbed into the liquid as tiny bubbles. Don't overdo this as too much sugar, or using weak or unsound bottles, could lead to the top blowing off or exploding bottles.
9. Drink!

Cider Vinegar

Cider vinegar is created by the action of acid bacteria on alcohol, and although these bacteria do occur naturally in the atmosphere, we found that simply

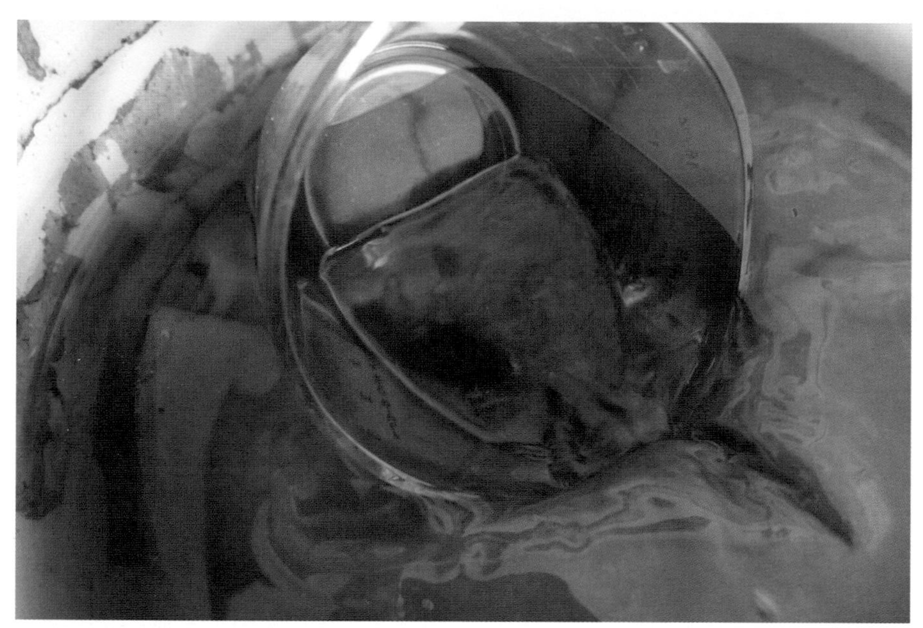

Vinegar mother.

leaving cider exposed to the air produced a number of interesting moulds but not vinegar.

To produce vinegar reliably you need to obtain some vinegar mother, a jelly-like cellulose substance that grows very well in the right environment – a vat of alcohol. Raw, unfiltered apple cider vinegar containing mother can be found at most health food, home-brewing or winemaking stores, or beg a fist-sized lump from a cider- or vinegar-making friend. Whilst not particularly appealing to look at and with the texture of a jellyfish wearing a rubber swimming cap, it is completely harmless. Put the vinegar mother in a plastic container, add cider, cover the open top with muslin that you tie in place, and leave for several months. And that's it.

The mother converts alcohol to vinegar at 1 per cent per week, so 7 per cent cider needs at least seven to eight weeks to convert. If you're fussy about bits of slimy mother being sprinkled on your fish and chips you can draw off the majority of the vinegar and boil it before putting it into a clean barrel or bottles (plastic caps only). Otherwise the mother can be filtered out using a coffee filter, or can be left in the vinegar and ignored (it will multiply over time).

Remember to add more cider to the remaining mother in the original vinegar tub to keep the process going. If you have surplus wine (raised eyebrows all round), have a separate barrel for that, add vinegar mother and create your own wine vinegar; and you can use beer to create malt vinegar. Vinegar's shelf life is almost indefinite: because of its acid nature, vinegar is self-preserving and does not need refrigeration.

RASPBERRY OR BLACKBERRY VINEGAR

This is my favourite salad dressing ingredient, and it is also perfect to put in a marinade with olive oil for lamb kebabs. Allow ½ltr of cider vinegar to each ½kg of berries. Put the fruit in a glass bowl. Add the vinegar to the fruit and leave for three to five days, stirring occasionally. Strain off the liquid, discard the fruit and measure the quantity of liquid. Add 85g white sugar per ½ltr of vinegar. Boil together for ten minutes, and bottle.

PRESERVING AND PROCESSING YOUR HARVEST

Chutneys and Relishes

When I was a child, chutneys and relishes were shop-bought ultra-processed concoctions spread inside a sandwich to give a much needed kick to a slice of bland cheese. These days they get used to complement rather better cheeses and as home-made burger toppings, stirred into a tomato sauce for adding complexity of flavour, as a dip for a cold or hot sausage, and lots more. My favourite chutney is made with Victoria plums and is best with strong, hard cheeses or spooned into barbecue sauce to give a sleek gloss to pork or beef ribs.

Sweet chilli jam is our most frequently used condiment. A gob of it stirred into cheese sauce makes a marvellously piquant macaroni cheese, while spooned into a bowl of mayonnaise it adds joy to cold fish and meat or potato salad. Putting a jar of this on to the table when serving meat or fish pasta recipes, baked potatoes, steak, frittata, or any dish where someone might ordinarily reach for a ketchup bottle, means I have to make gigantic quantities if it's to last until it's time to make it again. If you have a greenhouse or polytunnel, grow lots of peppers, chillies and garlic, and you'll have plenty enough for your purposes.

Fermented Food

Beyond wine, beer, cheese and sourdough, from kimchee to sauerkraut, fermented foods are the newest in a long-line of superfoods. Whether it's kefir cheese using surplus milk, or kombucha tea, you can ferment veg, fruit, milk and more. It's not only fans who swear by fermented foods for improving gut health – doctors are now also supporting this claim. A smallholder with surplus ingredients is in a good position to experiment; just make sure you have plenty of glass jars, fridge space and shelving.

Jams and Jellies

As with chutneys, there are many books on jam and jelly making, but there are some smallholder favourites to make from orchard and foraged fruits, including crab apple jelly. If you're lucky enough to have a quince tree, you are blessed. If not, plant one now; the fruits are deliciously scented. Membrillo is a thick, sliceable quince jam traditionally served with hard cheeses, but it is equally delectable with poultry and meat, either spooned on the side of the plate or melted into a meaty gravy. When making membrillo you will also have gorgeous liquid to make quince jelly.

Freezing

Is anything easier than freezing? You bag or box something up, stow it in the chilly depths, and weeks, months or years later, out it comes to grace your plate. Meat officially has a twelve-month frozen 'use by' date, but we've happily eaten goose, duck, beef and lamb that has languished forgotten in the bottom of a chest freezer for many years, and neither it nor we have suffered for it (although it wouldn't be suitable for sale).

As well as meat and poultry, freeze blackberries and chillies fresh off the bush, and so much more. All my old preserving books say you should blanch peas before freezing but it's not necessary, although broad beans benefit from a brief dousing in boiling water, cooling in cold water, and thoroughly draining before freezing. Whenever possible avoid processing to keep things quick and easy: cut plums in half, stone and freeze on a tray, and then a day later tumble into bags and keep in the freezer until needed for cakes and puddings or sauce to adorn roast duck.

Freezers mean you can create your own healthy, delicious, additive-free ready meals. Dishes such as beef casseroles, steak and kidney pie filling, red cabbage casserole, mutton tagine, curries, veg and beef chilli and ratatouille can all be cooked in large batches and frozen in portions.

Drying

It's definitely worth growing some beans to harvest their seeds. Simply leave them on the plant until the outer pod is all withered, then pick and pod. Put the bean seeds on baking trays to dry out in a warm oven or an airing cupboard until hard, and then store in a sealable container until needed. We do this with gloriously pretty borlotti beans, but you can do it with haricots, flageolet, butterbeans, cannellini, edamame, kidney beans (from runner bean plants) and others.

Because dried beans contain lectin, which can give you stomach ache, all varieties should be soaked for at least eight hours, and boiled in fresh water for ten minutes; then they are ready to cook until soft in your stews, ratatouilles and other favourite dishes.

Charcuterie

Growing up with European parents, charcuterie was always part of our diet, and there was nothing exceptional about eating salami, air-dried ham and other cold meats for breakfast. These are high value products, so it is well worth making your own, and you can adjust the ingredients to suit you. Not keen on fennel? Leave it out. Appreciate a really spicy hit? Add more chilli and garlic. Prefer smoked paprika to sweet? You know what to do. You can buy sausage making and charcuterie kits that include everything you need, bar the meat; or you can individually source natural casings and all the flavourings that appeal.

If you don't have an airy, dry, fly-free place to hang your charcuterie while it's developing, create a meat safe from timber and flyscreen materials or muslin that can be hung from a rafter in a cool garage or shed.

Smoking

From cheese to fish to meat to vegetables, the gourmand smallholder with DIY skills can easily build their own hot and cold smokers. You can make hot

Ratatouille.

Borlotti beans.

smokers from all sorts of old metal containers, from bread bins, cake or biscuit tins; however, a cold smoker takes a bit more effort unless you have an open fireplace and are happy to hang your food over the embers.

Sausages

Sausages must be one of the most popular foods there is. Making your own is highly recommended and an absolute pleasure to do with family or friends. Decide on how much of the fat you want to incorporate into the sausage and mince that, too; a sausage doesn't work without fat, so aim for a minimum of 15 to 20 per cent for a good British banger, although 30 per cent or more is common. Add seasoning and flavourings to a fistful of the meat and fry it: if it tastes great you have your recipe, if not, adjust, and fry another gobbet until it tastes delicious.

You can get a dedicated sausage maker, but we use an ancient food mixer with a mincer attachment and sausage-making nozzle. Buy natural casings, and soak and rinse them thoroughly as they are preserved in salt. Check out videos to get to grips with the technique for twisting into links – or just twist away and have fun.

THE DELIGHTS OF OFFAL

We're a bit weird about offal in the West. From sneering at tripe, gagging at brains, having ghastly memories of school dinner liver and being odd about tongue, we don't give it much of a chance. This is a huge mistake. Not every type of offal will do it for you, but there are so many delicious possibilities that you should at least give some of it a try. The pickiest eater will probably find chicken liver pâté acceptable, and this can be used as an introduction to the more eyebrow-raising delicacies.

Oxtail, cooked long and slow with leeks, carrots and celery from the garden, plus thyme and slugs of wine or beer, is simply glorious: a winter feast that will have the most finickity picking up the bones to

Making Sausages

Deboning the pork	
Diced ready for mincing	
Discarded bits – use for stock	
Cumberland sausage	
Sausage links	

curl a tongue round every last morsel – this is not dainty grub.

Another delicacy is lambs' kidneys cooked with softened onions until they are no longer bloody but still tender, with a splash of marsala or sherry and a gob of French mustard. It's a hardship that alongside 12–14kg of lamb you still only get a pair of tiny kidneys, so these are treats to be savoured. Luckily, beef kidneys are on a grander scale and will provide the essential loveliness in a steak and kidney pie or pudding.

DAIRY PRODUCTS

Cheese

Where to start? There are encyclopaedias on this topic worthy of consulting, but if you have milk you can make the simplest of cheeses using some cheese or yogurt starter (C20G starter is readily available on-line), a cooking thermometer, butter muslin and 4ltr of fresh milk left to stand at room temperature for an hour.

In a stainless-steel saucepan heat the milk slowly to 86°C, stirring while it warms; when up to temperature take it off the heat and sprinkle $^1/_8$ teaspoon (this equates to a pinch, or 0.625g) of the starter to the milk. Leave for five minutes and then whisk for twenty seconds to disperse the starter evenly. Cover the pan and leave at room temperature for twelve hours.

Put a stainless-steel colander into a large bowl and line it with damp butter muslin, ladle the curds from the pan into the colander and let them drain for two hours at room temperature. Add sea salt (about a teaspoon) and stir it into the curds. Tie the ends of the muslin into a knot and suspend over the pan (a wooden spoon handle pushed under the knot works for this). Drain the curds for six to twelve hours; the longer you drain it, the firmer the cheese.

And that's it – eat it on oat cakes or any way you fancy.

From here you can progress to other fresh cheeses, and soft, semi-soft and ripened harder cheeses of every kind. Cheese making is a special sort of alchemy that requires strict hygiene and precision in timing and temperature control, but can still be achieved in the kitchen.

Cream

When separating cream from the milk, skimmed milk will be the useful by-product. You can use an electric cream separator, or simply stand the milk in jars in the fridge so the cream rises to the top to be skimmed off with a spoon, or use an old-fashioned separator dish.

Cheeses ripening in the fridge.

Cream separator.

Butter

Ten litres of milk provide around a litre of double cream (this varies depending on the breed of cow or goat), and 1 litre of double cream will produce 225g of butter. Once you have your double cream, add salt if desired and whisk beyond the whipped cream stage until the butter forms. Rinse the butter in cold water until the water runs clear, then using spatulas or butter hands (also called butter paddles or Scotch hands), pat it into your preferred shape, removing air bubbles and more of the moisture as you go. Keep the butter in the fridge where it will firm up, or store in a freezer in useful-sized lumps. More importantly, spread it on some fantastic bread or toast and eat it.

Ice Cream, Soap and More

A quick note about making things that you really like to eat. What's the point of making cottage cheese if you hate it? Try instead thick yogurt garnished with fresh strawberries and honey all from the holding that will transport you to the Greek islands. Home-made raspberry ripple ice cream will make you wonder why you ever bought commercial alternatives, and will preserve summer memories in the freezer. Clotted cream and fudge aren't just for holidays when the milk is building up in the fridge.

And what about non-foodstuffs? Soaps, shampoos, cleansers and moisturisers can all be made with surplus milk and exactly to a recipe that suits you. Be aware that there can be hazards in soap making if the processes are not properly followed – very high heats, plus explosion potential if you add water to the lye (sodium hydroxide) powder rather than vice versa – so it's imperative to think about safety before you get making, and to ensure that young children and anyone not involved in the process are kept well out of the way.

SKINS

We take our skins to a tannery to produce beautiful sheepskins, and these are an important strand of our income. If you have just a few skins each year you can have a go at tanning them yourself.

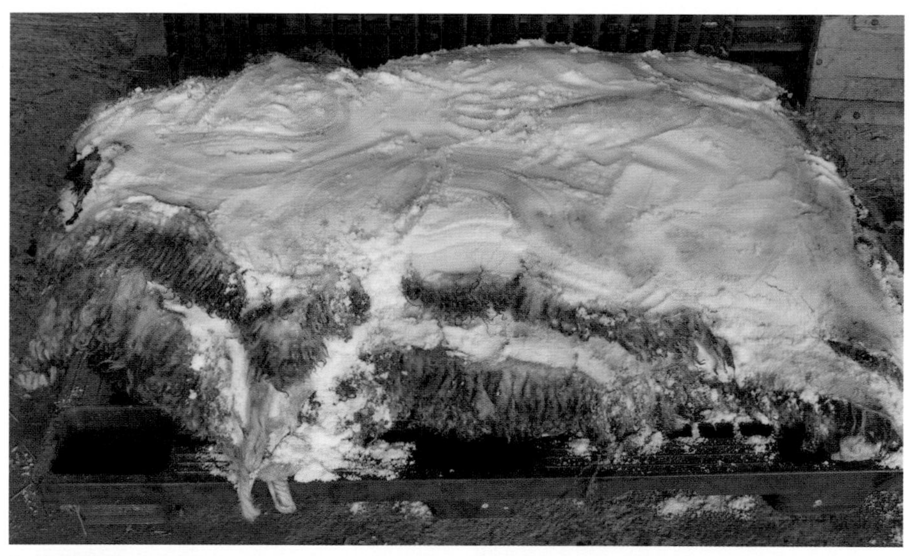

Freshly salted skins.

Herdwick lambskin – the finished product.

Tanning Your Own Skins

Ensure suitable gloves are worn throughout the process as bacteria can be present in the early days and the salt alum mix isn't pleasant on your own skin – which doesn't want to be cured.

Skinning: If home butchering you'll have immediate access to the skin; if collecting from the abattoir, ask them to do a preliminary salting and collect within twenty-four hours of slaughter.

Preparing the skin: Use a sharp knife to scrape away the fat, being careful not to cut the skin. Use a fine white salt, such as industrial dried vacuum salt, to salt fresh skins quickly or they start to decompose and the hair starts to slip. Lay the skin in a cool place with the fibre side down and apply the salt in a generous layer (about a centimetre) all over the flesh side; leave it to draw moisture for thirty-six to forty-eight hours. Pools of liquid will drain off, so this is an outside/shed/barn job. If you salt skins on a pallet make sure it has no iron nails or staples as this causes discoloration.

Tanning: When you're ready to start tanning, rinse the salt off the skin. In a suitable-sized bucket mix equal amounts of salt and alum (aluminium potassium sulphate) in warm water, about 25g of each per 1ltr of water. Leave the skin in the bucket for two weeks, stirring daily; you may need to change the liquid halfway through for larger skins. You are looking for the skin side to whiten.

Drying: Once the skin has whitened, remove it from the bucket and wash thoroughly with warm water (you can add a little detergent if you wish, but rinse it out well). Now the skin needs to be dried. Skins tend to shrink and curl, so tack it at the edges on to a piece of ply to stretch it out. Depending on the size and thickness of the skin and the weather, it can take up to a month to dry fully.

Finishing: You can use sandpaper to remove any remaining fat or ugly skin tissue from the inside. Use a brush to comb through the fibre/wool, and your skin is ready to use.

GENERATE YOUR OWN ENERGY WITH NATURAL RESOURCES

Firewood

If you have woodland or hedgerows, managing them effectively can provide all the fuel you need to heat your home, and potentially cook with it too, using a solid fuel stove. We have an 11kW woodburner with boiler that heats radiators throughout most of the

Firewood.

Solar panels on the cowshed roof.

house very effectively, fuelled entirely from our coppicing and hedge-laying efforts.

If using a chainsaw, wear all the appropriate protective gear, and make sure you have had the relevant training. Season logs at the rate of 2.5cm per year. Moisture levels should be 20 per cent or lower before burning – and sweep those chimneys every year before lighting a fire.

Solar, Wind and Water Power

Solar power can be used for everything, from keeping your electric fencing going, to making a significant contribution to household electricity usage. Whether you have a substantial solar array on the roof of a barn or house, or small mobile panels for individual pieces of equipment, the potential for capturing solar power is increasing and the costs are going down as the technology continues to improve. We have enjoyed free hot water from solar water-heating panels, and electricity from the photovoltaic solar panels on our cow shed.

Domestic wind turbines can't just be perched on any windy spot to produce clean electricity. The choice of specific location is critical, and not all sites that are on top of a hill, for example, are suitable. Living by a river may give you options for capturing power through a hydroelectric system. If you are considering solar, wind or hydro installations, do get recommendations to find reputable advisers and installers. Being entirely off-grid no longer means mimicking Little House on the Prairie: with back-up from a generator when access to renewable energy is low, modern life continues apace.

Spring Water, Wells and Boreholes

We have spring-fed water and wells, and these supply half the water troughs on the farm. The water tends to dry up at the height of summer, but a simple switch in the house means we can swap to mains water until enough rain has fallen to replenish the free supply. It's particularly important on a smallholding with water pipe and livestock drinkers to keep an eye on the water meter to spot and cure any leaks, as these can prove incredibly expensive, as well as being a waste.

If you want to construct a borehole or well to abstract water, contact the Environment Agency as consent may be required. Siting and constructing boreholes are specialist tasks, and advice should be sought from a hydrogeologist. The Environmental Health Department in your local authority can test these water supplies to ensure they are fit for consumption.

18 Smallholder Worries

I'm no counsellor so can't anticipate or resolve all smallholder concerns, but there are some constants in the questions I get asked. One of the common ones is 'What do you regret most?', which is not so much a smallholding issue as an approach to life's challenge. I don't really regret anything on the smallholding front, except perhaps not doing some things sooner, simply because if something wasn't to our liking, we stopped doing it and tried something else.

A good example is when we started our sheepkeeping with Jacob sheep, described ubiquitously as the smallholder's sheep. That may be so, but we had the four-horned variety and they were wild and nothing but trouble. After a couple of years we sold them and bought some Welsh Mountain Badgerfaces instead; we still keep them thirty years later. I frequently tell people that livestock are not their children: if they don't suit you, mark them up and sell them on to someone who will take joy in them, and get something else before the experience ruins for you not just a breed but a whole species.

FREQUENTLY ASKED QUESTIONS

Can I Ever Go On Holiday?

This question always pops up from participants in our Introduction to Smallholding course, and is normally asked with a hint of panic or despair. It's not as if you can put a flock of hens, three pigs and a small herd of goats in kennels, alongside your dog and cat, is it?

Holidays have huge significance when you spend the whole day in an office, and the prospect of foregoing a much-longed for break can be almost unthinkable. In reality, many people start smallholding to give themselves a complete or partial break from the office grind, which might in itself offer a sense of semi-permanent holidaying. It was nine years after moving to our farm in Devon before we took a holiday. Here we were, in a beautiful, quiet piece of glorious countryside – it was exactly the sort of place we used to choose to go on holiday, only we were living here full time.

Then we realised that it is an important part of maintaining sanity to have a break, and now we do holiday and enjoy every precious moment away. I can reassure you that there is a variety of options that will still let you visit relations in New Zealand, take a week rattling round France, or go diving in Hawaii, funds allowing. We choose to admire sheep in Cumbria or cows in Cornwall: each to their own.

If you are smallholding, it's highly likely that there are other smallholders living near you. You might be able to come to a mutually beneficial arrangement not unlike a babysitting group, which means that you all pitch in and offer your services when one of you wants or needs to get away for a bit. This will require some shared planning so that you don't all book to go away at the same time. If you have a few hens and not much else, even a non-smallholding neighbour may find the prospect of feeding and watering your birds a thrill, particularly with warm fresh eggs on offer – you can but ask.

On our 3.5-acre smallholding we were inundated with people asking if they could holiday-sit for us. At the time we had a couple of pigs, a few sheep, hens, ducks and geese, dogs and cats. Young and not so young families asked up to a year in advance if they could *please* come and live at our place for a week or two. This was a wonderful arrangement – everyone was happy. We got to go away, while our friends took a country break and had the fun of feeling 'hands on' without too much responsibility.

As livestock numbers increased and cows came into the picture, we needed someone with experience looking after things. Family or friends come and housesit and take care of cats and dogs (and the poultry if they wish to), and we pay someone diligent and trustworthy to come morning and afternoon to do all sheep, pig and cattle chores. There are plenty of smallholding 'sitters' around, so seek out recommendations. Some will come daily, others prefer to live in – decide what arrangement works best for you. Agricultural or vet students make great sitters: rarely daunted by whatever faces them, they get stuck in and deal with the day-to-day, besides any unforeseen emergency. An active but retired smallholder who really understands the management tasks would also be a good choice.

Make sure you leave clear, precise notes, as much feedstuffs and other livestock consumables as they could possibly need, a float for unforeseen expenditure, contact details for the vet and any local farmer/smallholder friend, and reassurance that if something dies, they must not panic (and preferably not tell you about it either – it will still be dead when you get back from holiday and there is absolutely nothing you can do about it anyway). Our animal-sitting notes include a farm map with names of fields; numbers of each type of animal and where they will be; frequency of any mucking out, how to do it, where the feed and implements are, where the muck should be put and where the clean straw is kept; and any minor issues that you are aware of and have treated so that they don't panic or think they've caused an issue.

We are clear about the times things need to happen – livestock love their routines and are far more amenable if these are maintained. Noting where to find stopcocks, fuses and trips will save frustration and worry. We make sure that all feed is clearly labelled – you really don't want pig food going to the hens or vice versa – and that the appropriately sized scoop is with the right foodstuff. To help avoid mishaps we pay our animal sitter for an extra morning and afternoon to do all the chores with us a day or two before we go away, which picks up on any forgotten detail.

What Happens If I'm Ill?

There are many solo smallholders out there, hard-working heroes all. If it's summer, with long dry days, accommodating ground, and the only thing that needs extra attention are the determined weeds in the veg patch, those of us who do our smallholding as a duo or family can quite easily accommodate one person down, whether it's a broken bone, a dose of summer flu, or time away from home attending to other business.

But sometimes the one person who can do a particular task is indisposed, so what do you do? We normally alternate our routine farm chores: one day I'll do the pigs, poultry and sheep, the next day the cows, so for the most part we are interchangeable. Although I'm happy to drive the tractor and trailer when haymaking or collecting logs, I am not, to my shame, a daily tractor user. I've had instruction on how to use the front loader to pick up and shift large bales of hay or straw (we're talking a quarter of a tonne here, so wheelbarrows are not an option) but it's not something I do as a matter of course. For emergencies I have written instructions in words I understand that I can refer to if needed. I also have incredibly helpful friends who will hitch up and empty the tipping trailer when it is full of muck.

Over an extended period it's simply not possible to do the work of two people if you're already stretched to your limit. There's a clear message here for smallholders doing this on their own: don't take on too much, or the pleasures can turn into pain. For everyone, you need a network of skilled friends and neighbours who can help out when you're in trouble, or details of professional contractors who will respond quickly and efficiently to your call for help (and some cash to pay for it).

Can I Combine Work With Smallholding?

There are very few smallholders who generate enough income from their holding to allow them to give up paid work entirely. If you are well organised, you can hold down demanding careers and manage a smallholding at the same time. For ten years, Andrew and I had full-time jobs that frequently required him to hop on a plane and go off for a week at a time, while I worked closer to home but also attended significant numbers of work events in the evening. The key was to have absolutely fox-proof poultry runs; gates, fencing and water supplies that were well maintained; and feed and other consumables always in stock. We took holiday at lambing time, and spent the weekends doing all the livestock chores and smallholding maintenance.

If your idea of relaxation is to come home from work, throw off the suit and pull on the boilersuit to clean out the hen hut and earth up the potatoes, then smallholding will certainly provide plenty of that. But if you regularly finish your work shift too exhausted to stand, then smallholding is probably not the answer for now.

What If It's All Too Much?

The temptation to try your hand at absolutely everything smallholding, all at once, can be overwhelming. But the best advice is to take it a little slowly. Get well acquainted with the mechanics of new projects – particularly those that are livestock related – before going entirely wild, ending up with a menagerie akin to a rescue centre, with more husbandry or medical needs than you are ready to deal with. As I said at the beginning of this book, this way of life is meant to bring you joy, not burden you with guilt, angst, aching bones and a broken spirit. If something doesn't work for you, or you simply don't like doing it, stop. Focus on the things that intrigue and inspire you, whether it's new or existing smallholding projects.

A smallholder's 'to do' list will be never ending, but don't let this run you ragged. Focus on the essentials, and if a single task seems overwhelming, divide it up into smaller, manageable chunks. It's amazing how a little progress gives you the boost you need to keep on rolling.

As I get older and the topic of retirement looms more frequently, I want to shape that into a continuingly active life phase – but there's no point pretending I will have the stamina or the will to be the same person I was at thirty or forty. Doing less of something, or stopping doing certain things altogether, is not about failure, in the same way that having fifty sheep is not necessarily more joyous than having ten. Take control of being the smallholder you want to be, and don't let others define what that should look like for you.

Do I Have To Do It All?

The question of things being too much links directly to that of feeling you have to do it all to be a 'real' smallholder. You absolutely do not have to do it all. You couldn't possibly do it all anyway. Doing it all would take a whole community of smallholders. Pick out the things that make sense for you. If you want to try lots of things before you decide on something more stable, you can have a go at all sorts of smallholder activities to truly test out what suits you best. You might be surprised by what appeals and what doesn't.

Will I Be Able to take My Livestock to the Abattoir?

Only you can answer this one. If you are a meat eater, buying it from other sources does not stop an animal from being killed in order to put protein on your plate – but there are some people who cannot, in the end, take their livestock to the abattoir. Have another read of Chapter 14, focus on the practicalities, and contemplate the exceptional meat that will result from your efforts.

KEEPING THINGS SECURE AND MANAGING PREDATION

Theft and Trespass

The saddest predation is that caused by human theft. Rural crime seems to be permanently on the increase, and ranges from the opportunistic thief to stealing to order. Whole flocks of sheep are rustled, while tractors, quad bikes and poultry are also stolen, with turkeys and geese particularly vulnerable at the lead-up to Christmas. So what can you do to keep your animate and inanimate possessions safe?

- Talk to the Police Rural Crime Team to get advice on what the particular risks are in your area, and guidance on how to secure your animals and possessions.
- Use technology to your advantage – video cameras can capture crucial evidence, and consider wildlife cameras with remote viewing. For those without an electricity supply, battery-powered wi-fi cameras are available that you can access from a smartphone.
- Put up signage that says you have CCTV, an alarm system, large guard dogs and so on.
- Put motion sensor lighting in vulnerable areas.
- If you work from home, don't ignore excessive bleating, mooing, quacking or honking during the day (or night). It's probable that something is going on that you should check out.
- Join your neighbourhood social media groups; police and others in the area can alert you to any current issues.
- Take out insurance cover.
- Keep tools and equipment secure, particularly those that could aid a break-in such as bolt croppers, ladders and crowbars.
- Avoid keeping precious livestock off site, and where possible don't keep your most desirable livestock at the boundaries of your holding.
- Take photographs of your livestock: these help in social media appeals if stock is taken.
- Paint or mark portable equipment with your postcode.
- Don't post photos of valuable equipment on social media, or tell the world you're going away for a week.

The Risk from Dogs and Dog Walkers

Dogs not kept under control by their owners can hunt and kill livestock from poultry to sheep, goats and even cattle. I wish there was a simple answer to this one, but if you have a footpath running through your holding, it's almost inevitable that you will have trouble with dogs. For some unfathomable reason, there are still dog owners who can't imagine their pooch as hunters with a killer instinct, and see the request for keeping them on a lead as an infringement of their rights. If you have dog walkers coming through your holding you may have to fence the path so that dogs cannot access your stock. Make sure any electrified fencing is labelled and doesn't obstruct footpaths.

CONTROLLING PESTS AND VERMIN

Rats

The bane of every smallholder's life, rats pose all sorts of risks to your livestock, contaminating and eating their feed, stealing eggs, killing birds, and spreading infections such as salmonella, leptospirosis, ringworm, E. coli and mycoplasma, besides being carriers of parasites including fleas and mites. For young birds kept in a covered run consider putting mesh on the bottom too, to give day-time protection from rats. Large rats will attack piglets, lambs and adult birds. There are various ways to control (no euphemisms intended, I mean kill) rats, but first and foremost you need to do everything you can to dissuade them from taking over (*see* panel). A pair of rats can become two hundred in just one year, so it doesn't take long for a few rats to become a major infestation.

Cat catching a rat.

MINIMISE THE PRESENCE OF RATS WHEREVER POSSIBLE

- Keep feed in rat-proof containers. This doesn't necessarily mean investing in expensive feed bins: metal dustbins and old chest freezers make excellent rat-proof feed bins.
- Clear up spilled feed, and always remove open feeders at night to a rat-proof container.
- Protect electric wiring in outbuildings in galvanised metal trunking: you don't want a rat starting a fire by eating through the wiring.
- For poultry:
    - Use treadle feeders.
    - Repair and block any holes into poultry huts, and use galvanised weldmesh (1 × 1cm squares maximum), not chicken wire, that a rat can't chew through.
    - A thick concrete pad underneath huts with an apron bigger than the hut will stop rats living below it. Use broken bottle glass below the concrete as a further deterrent. If you build huts on stilts, make sure there is enough clearance for you to access underneath it, rather than creating a perfect rat den.
    - Collect eggs before rats have a chance to gorge on them.
    - Screw metal sheeting to the bottom of hut doors, which will stop them gnawing entry holes.
    - Maintain a minimum 1m-wide space around the perimeter of the hut that is free of long grasses and weeds so you can check for rat runs and any rodent activity.

There are various ways to kill rats effectively, but some, such as poison, have serious downsides in that they impact on non-target species, even when you use proper bait boxes. The last thing you want is to have poisoned, dying rats picked up as food by owls and cats, or poisoned mice being eaten by your birds or pigs. The table describes other humane options that are open to you.

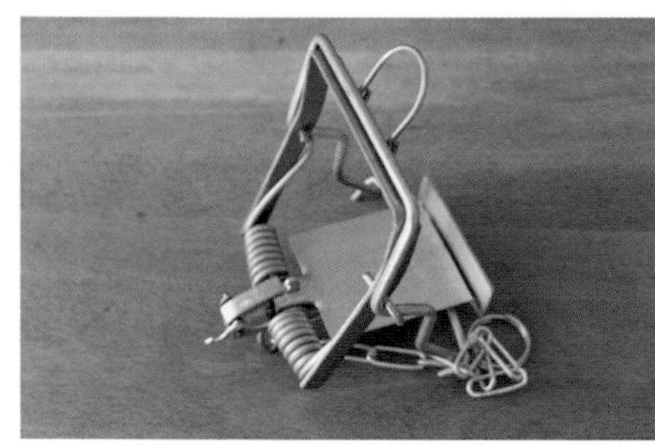

Fenn trap.

Methods of Rat Control

Methods of rat control	Advice	Pros	Cons
Break-back or snap trap	Place around the edges of a space, in a tunnel structure (for darkness and to stop your birds, pets, non-target species reaching inside). Bait with peanut butter.	Cheap, easy to set. Good for young, small rats. Can be left permanently in place.	Doesn't always kill – the rat can drag it around if partially trapped in it. You have to handle the rat to remove from the trap.
Fenn trap	A strong metal spring trap. Must be placed in a tunnel, as above. Bait with peanut butter. Can be secured to stop a rat dragging it away.	Highly effective, especially against large rats. Can be left permanently in place.	Be very careful when setting; they are strong enough to break your fingers. You have to handle the rat to remove it from the trap.
Terriers	Either your own or ratters with dogs will come to you. Particularly fruitful when an area is being cleared of rubbish or undergrowth.	Quick, effective, kills large numbers at one time.	Not a one-off or infallible (but then neither are the others). Can only catch those they can get at.
Humane or live cage traps	Catches rats live.	No harm to non-target species.	You now have to dispatch your rat. Releasing it somewhere else is simply relocating the problem and it may travel up to 5 miles to get back home.
Airgun	You have to be both careful and able. Use lead-free shot.	Effective if you are a good shot.	It can take patience, and may be time-consuming. You only see one in ten.
Battery-run traps	Electric shock kills the rat.	Effective.	Not the cheapest option.
Cats	Consider giving a home to a few feral cats.	Some cats are great ratters, others less so.	Your wild bird population (especially fledglings) will suffer.
Gas canister-powered automatic rat trap	Get a personal recommendation before you invest.	Automatic reloading after a rodent kill, no poisons involved.	Very expensive; opinions differ as to their effectiveness.
Noise emitter	Ultrasonic sounds can be heard by pets and some people, not just rodents.	Intended to keep rodents away from an area, therefore humane.	Not seen as particularly effective.
Sticky boards/ glue traps	Place where your birds (and wild birds) cannot get stuck.	Cheap, non-toxic, effective if baited with desirable foodstuff.	Inhumane and very much seen as a last resort even by the manufacturers. Check twice a day and any rats caught live still need to be dispatched (by you). Indiscriminate in what they catch.

Corvids and Other Predatory Birds

Corvids (crows, jays, rooks, magpies, jackdaws) can cause various problems for poultry, sheep, goats and pigs, stealing eggs and young birds, pecking out the eyes, udder and belly of poorly/cast sheep and goats, persistently attacking kids and lambs, and pecking holes into the backs of adult pigs. A Larsen trap, a type of cage trap, is designed to catch birds alive and unharmed, and they are used for the most part to capture magpies that then need to be humanely dispatched. A Larsen trap can be baited with food or with a live decoy magpie, provided welfare regulations are met. In Scotland, a cage trap must have an identifying tag obtainable from the police Wildlife Crime Officer. It is illegal to kill ravens in the UK.

Mink

American mink is widespread in the UK although I've yet to see one. Their preferred food is small mammals, including voles and rabbits, and they are also a risk to poultry and eggs, being aggressive predators. The Wildlife Trust recommends using live capture traps (cage traps) as the acceptable way of controlling mink, followed by licensed shooting. Traps must be fitted with an otter exclusion guard to avoid the inadvertent capture of young otters.

Foxes and Badgers

Far more controversial a subject than dealing with rats or the occasional magpie is managing foxes and badgers. Badgers are a protected species and it is illegal to trap or kill them for the sake of protecting livestock. If foxes are thriving in your area and encroaching regularly, even where your poultry are well contained, you might decide to keep numbers down by having them shot humanely by someone trustworthy with a firearms licence, using appropriate ammunition. Both species are powerful diggers, so if you have any option at all, site your poultry well away from any badger sett or fox earth. They may be nocturnal, but there are countless incidences of birds being predated by day.

If making hay in areas with known badger activity, avoid feeding this too soon to cattle because of the potential TB risk. The bovine TB bacteria can survive in hay just for a few days in warm weather, but up to forty days in cooler conditions in the autumn and winter. Research indicates that wrapped haylage that was contaminated with the bacteria when it

Fox.

Badger.

Guard llama with his flock.

was made is probably safe to feed to cattle after a month, although a small risk may remain.

You may enjoy keeping a guard llama or alpacas; they have been used effectively for many years to keep predators away from all types of livestock.

CATTLE AND BOVINE TB

All potential cattle keepers must be aware of the threat of bovine TB affecting a herd, no matter how small: it is something you will have to accept if you do decide to have cows on your smallholding. There are some practical things you can do to minimise the risk, although it cannot be eradicated:

- Ensure badgers are kept out of any building where you are housing cows, and that their feed is neither stored nor distributed where badgers can access (solid sheeted gates hung close to the ground leaving no more than a 7.5cm gap).
- Minimise wildlife access to feed, feed and water troughs, and mineral licks. Raise these off the ground where possible.
- Check feeders and troughs for wildlife faeces, and clean them regularly.
- Avoid mowing for forage any areas used by badgers as a latrine.
- Boundary fencing should prevent your cattle having nose to nose contact with neighbouring herds. Do not share water sources with neighbouring herds.
- Remove fallen stock and any dead wildlife speedily.
- Store manure for at least six months before spreading. Only spread manure from your own cattle on your land, and use disinfected muck spreaders to do it (or have one for your sole use). Spread manure on fields at least two months before they are used for grazing.
- Check the TB status and history of the source herd.
- Where possible breed your own replacements and use artificial insemination.
- Use wildlife cameras to identify badger activity in cow sheds and hay/straw stores.

Be aware that although incidence is low, alpacas, llamas, sheep, goats, pigs, cats and dogs can be infected by TB, as can humans.

BIOSECURITY

Keeping things healthy means maintaining a certain level of hygiene. No one expects animal housing or fields to be as clean as a hospital operating theatre or even your kitchen, but neither should it be an area of slops and faeces that welcomes maggots, flies and disease. Ask visitors visiting your livestock to dip their boots in a farm disinfectant, and follow quarantine protocols as described in Chapter 13.

There will be times when there are specific heightened bio-security risks, such as an avian flu outbreak, at which time the government requires additional protective measures for flocks of all sizes. Be aware of the measures likely to be put in place so that you can accommodate them with only minor tweaks. With every ensuing outbreak and subsequent lockdown, guidance is refreshed and updated, so do check current requirements.

USING A CONTRACTOR

What are the things on a smallholding that are really better carried out by the experts? Whether it's the need for expertise, brawn, or access to costly equipment, these tasks are things that you may not wish to do yourself; they include shearing, hedge trimming and haymaking.

Shearing

I have shorn very few sheep in my time, simply because it is the most back-breaking farm task I can think of, and it requires a very high level of skill. Also, the longer it takes, the more stressed the sheep become, so employing a professional means that it is quick and stress free, and the resulting fleece is of some practical use rather than a heap of sweaty fluff and fibre waste. Our shearer comes three times

Boot dip for visitors.

a year: mid-June to shear the adult ewes and rams; early September to shear any lambs we are growing on over the winter and beyond; and mid-February to crutch the ewes (removing the fleece from the tail and bum for cleanliness and management at lambing). We do have our own shearing kit and can do an emergency clip if it's needed, but mostly our equipment is used to trim the belly fleece from the lambs before they go off to the abattoir, a relatively simple job.

Hedge Trimming

Great swathes of farmed countryside in the UK are divided by living hedges, and a smallholding is as likely as a farm to have its share of hedges. They need managing, and our hedges are trimmed annually on a rotational basis (each gets a trim every three to five years) with a flail cutter. This is not a regular smallholder piece of kit, not least because it needs a very powerful tractor to run it, and it is an item used once a year, or less, if you have just the one boundary. When the trees and hedges are dormant and there's nothing much left to forage for ourselves or the wildlife, we call on a contractor to come and do the job.

Haymaking

Unless you have a passion for making your own hay and are happy to buy a fairly costly range of kit, it makes sense either to get a contractor to do this for you, or - even less stressful - to buy in the hay you need.

Niche Tasks

Depending on what sort of things you do on your smallholding, there may be a range of niche, specialist tasks that you might feel only a professional can do. In our case this is the annual hoof trimming of our cattle, and the seasonal artificial insemination (AI) of the breeding cows (two very different tasks carried out by two different specialist contractors).

For other smallholders, jobs in this category might include shearing alpacas, or trimming their feet; muck spreading (a couple of cows will produce more muck, if they overwinter inside, than you might fancy spreading on the land using a wheelbarrow and muck fork); and weed control. Then there are the desk-based tasks that you might be brilliant at, or for which you would welcome some specialist help: for example book keeping, building a website, and logo design/branding.

A PHILOSOPHY FOR LIFE

Perhaps it's a state of mind, or a philosophy for life, but spending hours and days of your time worrying about something you don't like doing is probably the antithesis of why you got into smallholding in the first place. If you come to a realisation that taking pigs to the abattoir, or growing Brussels sprouts, or dealing with a permanently aggressive cockerel is not how you want to spend your precious, all-too-short life, then do something practical about it. You can always buy your pork from another smallholder, plant purple-sprouting broccoli or French beans, and eat the confounded cockerel.

Glossary

A

abscess Localised collection of pus.
acidosis Condition in which the ph of the rumen is abnormally low <5.5, normally due to excess grain being consumed.
adoption *See* **fostering**.
afterbirth Placenta and foetal membranes expelled from the uterus after giving birth.
anaemia Lower than normal number of red blood cells.
anthelmintic Drug that kills certain types of intestinal worms, also known as wormer.
antibiotic Drug that inhibits the growth of/destroys micro-organisms/bacteria.
artificial insemination (**AI**) Placing semen into the uterus by artificial means.
auto-sexing Gender distinguished by colour.

B

baconer Pig reared for bacon slaughtered at 80–100kg at around 8–10 months.
banding (or **ringing**) Applying a rubber ring to the scrotum for castration.
bantam Mini version of larger poultry breed, about a quarter of the size.
battery hen Commercial hybrid laying hens frequently rehomed once past their production peak at around eighteen months.
biophilia The innate human instinct to connect with nature.
bloat Excessive accumulation of gases in the rumen of an animal.
boar Male pig.
body condition score (**BCS**) Numeric value that assesses the degree of fatness and condition of an animal's body.
bottle jaw Oedema or fluid accumulation under the jaw. Often a sign of infection with haemonchosis, also known as Johne's disease (*see* glossary entry below).
breech Delivery position in which the kid is presented backwards with its rear legs tucked underneath.
breed Group of animals with similar characteristics distinguishing them from other animals, and that are passed from the parents to the offspring.
broker/broken-mouthed Goat or sheep that has lost or broken some of its incisor teeth, usually due to age.
broody Female bird that stops laying and wants to sit on eggs and rear young.
browse Bushy or woody plants consumed by goats.
buck Sexually mature intact/entire (uncastrated) male goat used for breeding.
buckling An entire sexually immature young male.
bull Entire male bovine of breeding age, normally over a year.
bull calf Entire young male bovine up to the stage of yearling.
bullock Mature castrated male cattle destined for meat.
burdizzo Bloodless castration method that involves crushing the blood vessels leading into the testicles.

C

cade lamb Regional term for orphan or bottle-fed lamb.
calf Bovine less than a year old.
caprine arthritis encephalitis (**CAE**) Infectious disease in goats that causes arthritis and inflammation in organs and joints, brain, spinal cord, lungs and udder. Currently incurable.
carcass/carcase Dressed body of a slaughtered animal (intestines, head, feet and skin removed).
caseous lymphadenitis (**CL**) Chronic bacterial disease characterised by abscess development in the lymph nodes and other tissues. Highly infectious.
cast Animal on its back unable to regain its footing.
closed herd No new animals are introduced into the herd, animals are not hired out or taken to shows.
clostridial diseases Potentially fatal infections caused by clostridia (soil-borne) bacteria.
cockerel Male chicken.
colostrum First milk produced by a female after giving birth. Rich in antibodies, it helps protect newborns against disease.
concentrates Hard feed, also known as cake, compound feed, grains etc.
conformation Combination of structural correctness and muscling of the animal, including its frame and shape.
CPH number County parish holding number, which registers land as being for agricultural use.
creep area Allowing young animals in areas with restricted access, while older livestock are kept out.
creep feed Small pellets of high-protein supplementary feed given to young animals.
cross-breed Animal whose parents are of two different breeds.
cross-grazing Using two or more species of animals on the same land, because they graze in different ways and benefit the sward.
crutching *See* **dagging**.
cud Food that is regurgitated by a ruminant to be chewed again.
cull Animal no longer suitable for breeding, which is sold for meat.
culling Slaughtering an unwanted animal.

D

dagging Also known as 'crutching': the removal of fibre from around the tail and between the rear legs of a goat or sheep.

dam Mother.

dehorning Removal of the horns of a mature animal.

diarrhoea Also known as 'scouring': an unusually loose or fast faecal excretion.

disbudding Removal of the horn buds to stop horns developing.

doe Adult, sexually mature female goat.

doeling Young female not yet sexually mature.

drake Male duck.

drench Orally administered liquid medicine (n); to administer a liquid medicine (v).

dressing, or **killing out percentage** Percentage of the live animal that ends up as carcase.

dry doe/ewe/cow/sow One that is barren or not pregnant.

dry matter Foodstuff that remains once water content is removed.

duckling Young duck.

dystocia Difficulty in giving birth or being born.

E

ear tag Method of identifying animals by using a plastic or metal tag placed in the ear of the animal.

elastrator Tool used to apply rubber rings to the scrotum for castration purposes.

ewe Female sheep of breeding age. May be qualified as maiden ewe (not yet bred), or ewe lamb up to one year.

external parasites Fleas, keds, lice, mites, nose-blot flies, and ticks that feed on body tissue such as blood, skin and hair, resulting in irritation, blood loss and disease.

F

faecal egg count (**FEC**) Process of assessing the level of parasite load in livestock based on the number and type of parasite eggs found in the faeces.

faeces Manure or excrement produced by an animal.

fallen stock Dead stock.

FAMACHA © Acronym for Faffa Malan Chart: a method of using the colour of the inner eye lid to determine the level of haemonchus contortus (also known as 'the barber's pole worm') infection in sheep and goats.

fat stock/finished stock Livestock ready for slaughter.

fertiliser Natural or synthetic soil improvers, which are spread on pasture to improve fertility.

finish/condition Amount of fat that covers the body.

flushing Increasing nutrition in the few weeks before mating to improve fertility, or in the period before birth to increase birthweight; sometimes referred to as 'steaming up'.

flystrike Where fly eggs have been laid on an animal, which hatch into maggots.

fodder crop Plant grown for animal feed.

food chain information Regulation requirements if you are intending to slaughter animals for human consumption.

footbath Long trough filled with a chemical preparation in which livestock stand, for protection from/treatment of hoof conditions.

foot rot Infectious pododermatitis, a painful, bacterial infection affecting sheep, goats and cattle.

forage Edible plant material used as livestock feed.

fostering (adopting) Encouraging an animal to accept a newborn from another female.

freemartin A female calf born with a male twin (usually infertile).

G

gander Male goose.

gestation Length of pregnancy.

gilt Female pig that has not yet had a litter.

goose Female goose.

gosling Young goose.

H

hay Fully dried preserved grass.

haylage Part-dried preserved grass.

heat Period when a female is fertile and receptive to the male; also known as 'oestrus'.

hectare Metric unit of area equal to 10,000sq m (100 × 100m), or 2.471 acres.

heifer Young female cow up to the birth of her first calf or in lactation following her first calving.

hen Female bird over one year old.

herd number Required in the UK for identifying livestock belonging to an individual herd.

hogg *See* **teg.**

hogget Castrated male sheep usually 10–14 months, or meat from a sheep between 12–24 months.

hot carcass weight Weight of a dressed carcass immediately after slaughter prior to the shrinkage that occurs in the chiller.

hypothermia Condition characterised by low body temperature.

I

immunity Natural or acquired resistance to specific diseases.

inbreeding Mating or crossing of closely related animals; sometimes referred to as 'line-breeding' when carried out to pass on or strengthen certain desirable traits.

in kid, in pig, in lamb, in calf Pregnant.

internal parasites Parasites located in the gastrointestinal system in animals.

intramuscular (IM) injection Given straight into a muscle.

intravenous (IV) injection Given directly into a vein.

J

Johne's disease (*mycobacterium paratuberculosis*) also known as bottle jaw Bacterial disease causing severe weight loss and diarrhoea. Not currently curable.

joint ill (navel ill; pyosepticaemia) In newborns, an inflammation of the joints caused by bacteria entering the body, normally via an untreated umbilical cord.

K

keet Young guinea fowl.

kid Young goat (n). To give birth to a kid (v). The meat from an animal less than one year old (n).

kidding Goat giving birth.

kidding percentage Number of kids (including multiple births) successfully reared in a herd compared with the number of does that have been mated.

L

lactation Production and secretion of milk; the period when the doe produces milk.

lairage Area of holding pens where livestock is held at the abattoir before slaughter.

lamb Young sheep with its dam, or up to five months. Qualified as ewe lamb or ram lamb.

lambing percentage Number of lambs (including multiple births) successfully reared in a flock compared with the number of ewes that have been mated.

liver fluke Small leaf-shaped organism that thrives on wet land. Causes liver damage and is fatal if not treated.

lungworms Roundworms found in the respiratory tract and lung tissue.

M

maiden Female of breeding age not yet mated.

maiden milker Doe that comes into milk without being bred.

mastitis Painful and potentially serious inflammation of the mammary glands due to bacterial infection.

mineral Inorganic group of nutrients, including elements such as calcium, phosphorus and copper.

moult Shedding of old feathers and growth of new ones once a year, usually in the autumn, at which point birds stop laying for two to three months.

movement licence Legal requirement for moving livestock on and off premises.

movement standstill Restrictions on the movement of livestock off premises.

mutton Meat from a goat or sheep aged two years or older.

O

oestrus Heat period, during which females are fertile and receptive to the male.

open Female that is not pregnant.

orf Virus that causes contagious ecthyma in sheep and goats; a zoonotic infection – one that can be passed to humans.

orphan Orphaned lamb or kid, or one rejected by its mother, or from a multiple birth where the dam has inadequate milk to feed all her offspring.

oxytocin Naturally secreted hormone that encourages the contraction of the uterine muscles during labour and milk let-down; a veterinary product used to stimulate contractions and help lactation.

P

passive immunity Acquiring protection against infectious disease from another animal – for example, when a newborn consumes antibody-rich colostrum from its mother.

pastern Lower part of the leg, above the hoof.

pasture-fed Entirely grass- and browse-fed livestock, either fresh or preserved.

pasteurisation Partial sterilisation to make milk safe for consumption and improve keeping quality.

pedigree Family tree showing the ancestry of a registered, pure-bred animal.

piglet Young pig still suckling from its dam.

pink eye Infectious keratoconjunctivitis, a condition in which the conjunctiva (the membranes lining the eyelids and covering the white part of the eye) become inflamed or infected.

pizzle Penis.

placenta Organ that protects and nourishes the foetus(es) whilst in the uterus.

pneumonia Inflammation of the lungs, caused by a bacterial or viral infection.

point of lay Pullet approaching the age of laying.

polled Without horns.

polyoestrus Able to breed all year round.

porker Pig reared to pork weight, normally around 60kg live weight, usually reached at 4–6 months depending on breed.

poult Young fowl.

pour-on Chemical preparation for the control of internal/external parasites, which is applied to the skin and is gradually absorbed; an alternative to injectable treatments.

predator Animal/bird that lives by hunting, killing and eating other species.

prolapse Interior organ pushed outside the body cavity (for example vaginal, uterine, rectal prolapse).

prolific (prolificacy) Highly productive in lambing, kidding, farrowing etc; fecund.

pullet Young female bird of a year or less.

pure-bred Not crossed with another breed.

Q

quarantine Confine and keep an animal away from the rest of the herd to prevent the spread of disease.

quarter The teated section of the udder (half the udder of goats and sheep, a quarter of the udder in cows).

R

raddle Coloured pigment applied to the male's brisket to mark the females he mates; a harness used to hold a raddle crayon.

ram Sexually mature intact/entire (uncastrated) male sheep used for breeding.

ringwomb Failure of the cervix to dilate sufficiently, causing delivery problems.

rose veal Typically meat from veal calves of six to eight months.

rotational grazing Organised system of moving stock from one grazing unit to another.

roughage High-fibre feed that is low in both digestible nutrients and energy (for example hay, straw, silage).

roundworm Unsegmented parasitic worms with elongated rounded bodies that are pointed at both ends.

ruminant Animal with a multiple-chambered stomach that is able to digest cellulose.

S

scab Irritating skin condition caused by parasitic mange mite, *Psoroptes ovis*.

scouring *See* **diarrhoea**.

scrapie Fatal, degenerative disease affecting the central nervous system of sheep and goats.

scrotum Pouch of skin containing the male's testicles.

scur(s) Small rudimentary horn.

semen Combination of sperm, seminal fluid and other male reproductive secretions.

sharps Needles, syringes, scalpel blades, and anything else that can puncture the skin.

shearing Removing the wool/fibre using mechanical clippers or hand shears.

shearling Regional term for sheep up to its first shearing.

silage Fodder prepared by storing and fermenting grass or other forage plants in wrapped bales or in a silo.

sire Father.

sow Female pig that has produced at least one litter.

stag Male turkey.

standing heat Period in which the female will stand still and accept the male for breeding.

staple Length of a lock of shorn fibre.

steer Castrated male bovine over one year of age.

stillborn Newborn that is delivered dead.

stocking density/stocking rate Relationship between the number of animals and an area of land.

store Weaned animal not ready for slaughter, which is kept for fattening.

straw Stems of cereals such as wheat, barley or oats that are cut and baled and used for fodder or bedding.

strip cup Cup covered with mesh to draw foremilk for examination to detect abnormalities.

strip grazing Controlling grazing by confining animals to specific areas of land (often using electric fencing) for short periods of time before moving them on to fresh ground.

stun To render unconscious, for example prior to slaughter.

subcutaneous injection Given under the skin, but not into the muscle; sometimes shortened to 'sub-Q' or 'SQ'.

supplement Feed/minerals designed to provide nutrients deficient in the animal's main diet.

sustainable farming Approach that uses on-farm resources efficiently, reduces demands on the environment, and may help rural communities.

sward The upper layer of soil, particularly when the ground is covered with grass.

T

tapeworm Ribbon-like parasitic flatworms found in the intestines.

teaser Male that has been vasectomised to prevent reproduction used to stimulate females for mating.

teg Regional term for fat lamb in second season. Also used for weaned lambs.

tup (*See* **ram**.) Also **tip**.

tupping Breeding/mating time for sheep.

twin kid/lamb disease Pregnancy toxaemia, a metabolic disease affecting very underweight or overweight does/ewes carrying multiple kids/lambs.

U

udder Milk-secreting organ.

urinary calculi Stones formed within the male urinary tract. Common in goats, it is caused primarily by an imbalance of dietary calcium and phosphorus.

uterus Organ in which the foetuses develop; the womb.

V

vaccine Injection given to improve resistance to/prevent disease.

W

WATOK – Welfare of Animals at the Time of Killing The regulations requiring a Certificate of Competence for poultry slaughtered on the holding that is for sale or use beyond the immediate family.

wattles/toggles/tassles Small fleshy appendage attached on or near the throat area of the goat and which serves no known function.

weaner Piglet that has been weaned from its dam and is now on solid foods.

weaning Process of taking young animals away from their source of milk.

wet adoption Covering a kid that is to be adopted, with birthing fluids from the doe.

wether Castrated male (goat/sheep).

withdrawal period After treatment with a medical product, the length of time that must be allowed to elapse before meat or milk is allowed into the human food chain.

wormer Commonly used term for anthelmintic, medication for killing intestinal worms

X

Y

yearling Animal between one and two years of age.

yield Amount of milk, meat, fibre or wool produced per animal.

Z

zero grazing System of growing fodder but not allowing livestock to graze it directly; instead, the crop is cut and taken to the animals.

zoonosis Disease or ailment that is zoonotic – that is, one that normally exists in animals, but can be passed to humans.

Further Sources of Information, Equipment, Reading and Courses

BEES

The British Beekeepers Association
W: www.bbka.org.uk

National Bee Supplies
T: 0344 326 2010
E: info@beekeeping.co.uk
W: https://beekeeping.co.uk

CAMELIDS

The British Alpaca Society
T: 01392 270421
E: info@bas-uk.com
W: https://bas-uk.com

The British Llama Society
T: 07890 025064
E: admin@britishllamasociety.co.uk
W: www.britishllamasociety.co.uk

CATTLE

National Beef Association
T: 01434 601005
E: info@nationalbeefassociation.com
W: https://www.nationalbeefassociation.com

Raw Milk Producers Association
W: https://rawmilkproducers.co.uk

British Cattle Veterinary Association
T: 01452 725 735
E: office@cattlevet.co.uk
W: www.bcva.org.uk

AHDB Dairy
W: https://ahdb.org.uk/dairy

COWS (Control of Worms Sustainably)
W: www.cattleparasites.org.uk

GOATS

British Goat Society
T: 01322 611767
E: admin@britishgoatsociety.com
W: www.britishgoatsociety.com

Goat Veterinary Society
T: 07734 458412
E: gvs.enquiries@gmail.com
W: www.goatvetsoc.co.uk

Milking Goat Association
T: 01454 436046
E: office@milkinggoat.org.uk
W: www.milkinggoat.org.uk

PIGS

British Pig Association
https://britishpigs.org.uk

AHDB Pork
T: 024 7669 2051
E: info@ahdb.org.uk
W: https://ahdb.org.uk/pork

POULTRY AND WATERFOWL

British Waterfowl Association
T: 01531 671250
E: secretary@waterfowl.org.uk
W: www.waterfowl.org.uk

Poultry Club of Great Britain
T: 01830 520856
E: info@poultryclub.org
W: www.poultryclub.org

Goose Club
T: 01437 563309
E: contact@gooseclub.org.uk
W: www.gooseclub.org.uk

RABBITS

The British Rabbit Council
T: 01636 676042
E: info@thebrc.org
W: https://thebritishrabbitcouncil.org

SHEEP

National Sheep Association
T: 01684 892661
E: enquiries@nationalsheep.org.uk
W: www.nationalsheep.org.uk

AHDB Beef and Lamb
W: https://ahdb.org.uk/beef-lamb

SCOPS (Sustainable Control of Parasites in Sheep)
W: www.scops.org.uk

GOVERNMENT AND OTHER HELPFUL ORGANISATIONS

Rural Payments Agency (England)
To register land and for grants.
T: 03000 200301
E: ruralpayments@defra.gov.uk
W: www.gov.uk/government/organisations/rural-payments-agency

Rural Payments Wales
T: 0300 062 5004
W: www.gov.wales/rural-payments-wales-rpw-online

Rural Payments and Services Scotland
W: www.ruralpayments.org/topics/contact-us

Department of Agriculture, Environment and Rural Affairs (Northern Ireland)
W: www.daera-ni.gov.uk/contact

Livestock Movements in England:

Livestock Information Service
Live for sheep, goats and deer, the service is intended to extend to pigs and cattle.
T: 0844 573 0137
E: support@livestockinformation.org.uk
W: https://livestockinformation.org.uk
Livestock movements in Wales by EIDCymru www.eidcymru.org
Livestock movements in Scotland by ScotEID www.scoteid.com
Livestock movements in Northern Ireland by the Animal and Public Health Information System (APHIS) www.daera-ni.gov.uk/aphis-online-support
Livestock movements in the Republic of Ireland www.gov.ie

British Cattle Movement Service (BCMS)
T: 0345 050 1234
E: bcmsenquiries@rpa.gov.uk
W: https://secure.services.defra.gov.uk/wps/portal/ctso

Pig Movements
T: 0844 335 8400
E: eaml2@ahdb.org.uk
W: www.eaml2.org.uk

Department for Environment, Food and Rural Affairs (DEFRA) and Animal and Plant Health Agency (APHA)
England: T: 03000 200 301
Wales: T: 0300 303 8268
Scotland: call the relevant field office:
Ayr T: 03000 600703 E: APHA.Scotland@apha.gov.uk
Galashiels T: 03000 600711 E: APHA.Scotland@apha.gov.uk
Inverness T: 03000 600709 E: APHA.Scotland@apha.gov.uk
Inverurie T: 03000 600708 E: APHA.Scotland@apha.gov.uk
Perth T: 03000 600704 E: APHA.Scotland@apha.gov.uk
In Northern Ireland Department of Agriculture, Environment and Rural Affairs (DAERA) www.daera-ni.gov.uk/contact Tel: 0300 200 7843 E: daera.helpline@daera-ni.gov.uk

Food Standards Agency (FSA) www.food.gov.uk
England: T: 0330 332 7149 E: helpline@food.gov.uk
Wales E: walesadminteam@food.gov.uk
Northern Ireland E: infofsani@food.gov.uk
Scotland: T: 01224 285100 E: enquiries@fss.scot
W: www.foodstandards.gov.sco

Farming and Wildlife Advisory Group (FWAG)
E: info@fwag.org.uk
W: www.fwag.org.uk

Humane Slaughter Association
T: 01582 831919
E: info@hsa.org.uk
W: www.hsa.org.uk

Moredun Research Institute
T: 0131 445 5111
E: info@moredun.org.uk
W: www.moredun.org.uk

National Animal Disease Information Service (NADIS)
T: 07771 190823
E: contact@nadis.org.uk
W: www.nadis.org.uk

National Fallen Stock Company
T: 01335 320014
E: member@nfsco.co.uk
W: www.nfsco.co.uk

Rare Breeds Survival Trust (RBST)
T: 024 7669 6551
E: enquiries@rbst.org.uk
W: www.rbst.org.uk

Transport Certificates
Training and testing for the City and Guild Certificate of Competence in the Transport of Animals by Road is available online. www.hushfarms.co.ukAnimalsInTransit-c-40.asp, and through some agricultural colleges

National Office of Animal Health (NOAH)
T: 0208 3673131
E: noah@noah.co.uk
W: www.noahcompendium.co.uk

REGENERATIVE AGRICULTURE

Pasture for Life
T: 0333 772 9853
E: info@pastureforlife.org
W: www.pastureforlife.org

EQUIPMENT

Brinsea
Incubators, hatchers and candlers.
T: 0345 226 0120
E: sales@brinsea.co.uk
W: https://brinsea.co.uk

Interhatch
T: 01246 264646
E: sales@interhatch.com
W: www.interhatch.com

Domestic Fowl Trust
Housing, birds and more.
T: 01789 850046
E: dft@domesticfowltrust.co.uk
W: www.domesticfowltrust.co.uk

Nestera
Recycled plastic bird housing.
T: 01963 371563
E: hello@nestera.eco
W: https://nestera.co.uk

Solway Recycling
Recycled plastic housing and more.
T: 01387 730 666
E: info@solwayrecycling.co.uk
W: www.solwayrecycling.co.uk

Sedgbeer Poultry Processing Equipment
T: 01761 420058
E: info@sedgbeer.co.uk
W: www.sedgbeer.co.uk

Leisure Heating
For ceramic heat emitters.
T: 0115 937 2727
W: www.leisureheating.co.uk

Goat Nutrition Ltd
T: 01233 770780
E: info@gnltd.co.uk
W: www.gnltd.co.uk

Frenchall Goats
T: 01638 750665
E: info@frenchall-goats.co.uk
W: www.frenchall-goats.co.uk

Homestead Farm Supplies
T: 01295 713188
W: www.homesteadfarmsupplies.co.uk

Dairy Spares
T: 01948 667676
E: info@dairyspares.co.uk
W: www.dairyspares.co.uk

Shearwell
Ear tags.
T: 01643 841611
E: sales@shearwell.co.uk
W: www.shearwell.co.uk

Dalton Tags
Ear tags and pasterns.
T: 01636 700990
E: sales@daltontags.co.uk
W: www.daltontags.co.uk

First Tunnels
Polytunnels.
T: 01282 601253
E: Sales@FirstTunnels.co.uk
W: www.firsttunnels.co.uk

Ferryman Polytunnels
T: 01363 84948
E: info@ferrymanpolytunnels.co.uk
W: https://ferrymanpolytunnels.co.uk/

Polycrub
W: www.polycrub.co.uk

Vigo
Brewing, juicing and cider-making equipment.
T: 01404 892 100
E: sales@vigoltd.com
W: www.vigoltd.com

TANNERIES

Bradford Hide Company
W: www.bradford-hide.co.uk

Devonia, Devon
T: 01364 643 355
E: admin@devoniasheepskinsandtannery.co.uk
W: www.devoniaproducts.co.uk

Organic Sheepskins
The only organic tannery in England (Dorset).
T: 01935 891204
E: mark@nevillefarm.co.uk
W: www.organicsheepskins.co.uk

Institute for Creative Leather Technologies, Northampton
Tans and processes a wide range of hides and skins in small batches.
W: www.northampton.ac.uk/info/200174/british-school-of-leather-technology

Skyeskyns
Isle of Skye.
T: 01470 592237
E: office@skyeskyns.co.uk
W: www.skyeskyns.co.uk

Welsh Organic Tannery
The only organic tannery in Wales.
T: 07966 470421
E: info@welshorganictannery.co.uk
W: www.welshorganictannery.co.uk

COURSES

Smallholder Training at South Yeo Farm West
Debbie Kingsley and Andrew Hubbard.
T: 01837 810569
E: debbie@smallholdertraining.co.uk
W: https://smallholdertraining.co.uk

Humble by Nature
T: 01600 714595
E: info@humblebynature.com
W: www.humblebynature.com

Perrys Field to Fork
Butchery courses.
T: 01785 851911
W: https://perrysfieldtofork.co.uk/product/goat-butchery-course-copy

Fielding Cottage
Cheese courses.
T: 01603 880685
E: sales@fieldingcottage.co.uk
W: www.fieldingcottage.co.uk/cheese/cheese-making-courses

Specialist Cheesemakers Association
T: 020 7253 2114
E: info@specialistcheesemakers.co.uk
W: www.specialistcheesemakers.co.uk/cheese-making-courses.aspx

Soap School
Soap making.
T: 01484 310014
E: sarah@soapschool.com
W: https://soapschool.com

Grazing School
Regenerative grazing.
T: 015394 37794
E: info@rootsofnature.co.uk
W: https://rootsofnature.co.uk/grazing-school

ARTIFICIAL INSEMINATION SERVICES

Goat Genetics (goats)
E: info@goatgenetics.com
W: www.goatgenetics.com/ai-artificial-insemination

Genus (cattle)
T: 01270 616681
E: cs@genusbreeding.co.uk
W: www.absglobal.com/uk/contact

Cogent (cattle)
T: 0800 783 7258
E: info@cogentuk.com
W: www.cogentuk.com

Deerpark (pigs)
T: 028 7938 6558
E: deerparkpigs@gmail.com
W: http://deerpark-pigs.com/deerpark-ai-centre

VOLUNTEERING

HelpX
W: www.helpx.net

Wwoof
W: https://wwoof.net

WorkAway
W: www.workaway.info

Volunteers Base
E: volunteersbase.com@gmail.com
W: www.volunteersbase.com

FURTHER READING

Some of these books are not that recent, and others are out of print and only available second hand, but they continue to be of particular interest to budding smallholders.

Acreman, David *Field to Farm* (Bulldozer Publishing, 2009).
Agro Business Consultants *The Agricultural Budgeting and Costing Book* (ABC Books, published bi-annually).
Ashton, Chris *Keeping Geese: Breeds and Management* (The Crowood Press, 2012).
Beeken, Laurence *Chicken Manual* (Haynes, 2010).
Biss, Kathy *Practical Cheesemaking* (The Crowood Press, 2005).
Bromage, Gina *Llamas and Alpacas – A Guide to Management* (The Crowood Press, 2015).
Caldwell, Gianaclis *The Small Scale Dairy* (Chelsea Green Publishing, 2014).
Dowding, Charles *Organic Gardening The Natural No-Dig Way* (Green Books, 2018).
Fairlie, Simon *Rural Planning Handbook for Low Impact Developers* (Land and Sky Press, 2018).
Grandin, Temple *Humane Livestock Handling* (Storey Publishing, 2008).
Grohman, Joann S. *Keeping A Family Cow* (Chelsea Green Publishing, 2013).
Harwood, David *The Veterinary Guide to Goat Health and Welfare* (The Crowood Press, 2019).
Kingsley, Debbie *Keeping Ducks and Geese – A Practical Guide* (The Crowood Press, 2021).
Kingsley, Debbie *Keeping Goats – A Practical Guide* (The Crowood Press, 2022).
Lea, Andrew *Craft Cider Making* (The Good Life Press, 2011).
Morgan, Sally and Stoddart, Kim *The Climate Change Garden* (Green Rocket Books, 2019).
Nix, John *Farm Management Pocketbook* (ABC Books, published annually).
Seymour, John *The Complete Book of Self-Sufficiency* (Dorling Kindersley, 1996).
Shankland, Liz *Pig Manual* (Haynes, 2011).
Smith Thomas, Heather *Essential Guide to Calving* (Storey Publishing, 2008).
Stromberg, Loyl *Sexing All Fowl, Baby Chicks, Game Birds, Cage Birds* (Stromberg Publishing Company, 2002).
Thear, Katie *Cheesemaking and Dairying* (Broad Leys, 2003).
Thear, Kate and Dr Fraser, Alistair *The Complete Book of Raising Livestock and Poultry* (Pan, 1988).
Thomas, Paul *Homemade Cheese: Artisan Cheesemaking Made Simple* (Lorenz Books, 2016).
Tyne, Tim *The Sheep Book for Smallholders* (The Good Life Press, 2009).
Tyne, Tim and Dot *Viable Self-Sufficiency* (Home Farmer, 2016).
Waring, Claire and Adrian *Bee Manual* (Haynes, 2013).
Watson, Chris *The Cattle Keeper's Veterinary Handbook* (The Crowood Press, 2009).
White, Mark *Pig Ailments* (The Crowood Press, 2005).
Winter, Agnes and Phythian, Clare *Sheep Health, Husbandry and Disease* (The Crowood Press, 2011).

Acknowledgements

This book has been in my head and heart for many years, and found its early form in the content of the hefty information packs written for all our smallholding and livestock courses. A particular thank-you to my husband Andrew Hubbard for his support and help; to John and Maggie Boswell for their early inspiration and lifelong friendship; and to Pat and Philip Hutton for their kindness, help and friendship. Sincere thanks also to everyone listed below: the time, photos and encouragement you gave so freely is much appreciated.

Fleur Ketley
Harriet Smith, Crediton Milling
Jane Ross, Wytsend Herd
Jo Seymour Tavernor
Ruth and Graham Eggins, Hillside Farm
Simon Emms
Zelda Lawrence-Curran

Image Credits

Alice Saffron Worby, Barleylands Farm Park
Andrew Griffiths
Andrew Hubbard
Barnacre Alpacas
Becci and Katie Bea Murphy
Brinsea Products Ltd
Chiltern Livestock and Country Girl Media
Chris Just BVSc MRCVS
Clifford Freeman
Clipex
Crediton Milling Ltd
Dalton Tags
Danielle Drew
David Bingham
Dewsburys Pork and Pig Stud
Donna Groves and Jake Bennett, The Emu Snack Shack
Durabo Rare Breeds Centre
English Goat Breeders Association
English Guernsey Cattle Society
Fleur Ketley, White Goose Flower Farm
Gam Farm Rare Breeds
GardenLife
Gloucestershire Old Spots Pig Breeders' Club
Harry Pennington
Humane Slaughter Association
Hywel Davies (Aman Texels)
Ianaré Sévi
Jan McCourt
Jo Hagyard
John Deere
Joyce Greenslade
Katie Mander, Holthall Balwens
La Buvette
Leamlady
Liane Hadden, Sacrewell Farm
Lorna Emms
Martha Roberts, The Decent Company
Meredydd Jones
Middle Farm, Shropshire
Nigel Edwards
Otterburn Mangalitza and Barbara Meyer Zu Altenschildesche
Paddy Zakaria
Polycrub and Ronnie Robertson
Professor Gareth F Bath, FAMACHA System Coordinator
Rachel Jones
Richard Freedman, Nestera
Sam Lawrence
Sasha Distan
Smiling Tree Farm
Sussex Cattle Society
Tanya Olver, Terras Farm
Tereza Fairbairn
The Chicken House Company
www.thechickenhousecompany.co.uk
Tredethy Kunekunes
Veronica and Jules Hartley
Westgate Laboratories
Will Oliver, Vinery Herd
Other images and diagrams by Debbie Kingsley

Cover images

Front cover: All images by the author, except for bottom-left image by Fleur Ketley.
Back cover: All images by the author, except for the middle image of geese and dogs in the garden by Fleur Ketley.

Index